The Liquid and Supercritical Fluid States of Matter

The Liquid and Supercritical Fluid States of Matter

John E. Proctor
With contribution from Helen E. Maynard-Casely

CRC Press
Taylor & Francis Group
Boca Raton London New York

CRC Press is an imprint of the
Taylor & Francis Group, an **informa** business

First edition published 2021
by CRC Press
6000 Broken Sound Parkway NW, Suite 300, Boca Raton, FL 33487-2742

and by CRC Press
2 Park Square, Milton Park, Abingdon, Oxon, OX14 4RN

First issued in paperback 2022

Publisher's Note
The publisher has gone to great lengths to ensure the quality of this reprint but points out that some imperfections in the original copies may be apparent.

**Visit the Taylor & Francis Web site at
http://www.taylorandfrancis.com**

**and the CRC Press Web site at
http://www.crcpress.com**

Library of Congress Cataloging-in-Publication Data

Names: Proctor, John Edward, author.
Title: The liquid and supercritical fluid states of matter / John E.
    Proctor ; with contribution from Helen E. Maynard-Casely.
Description: First edition. | Boca Raton : CRC Press, 2021. | Includes
    bibliographical references and index.
Identifiers: LCCN 2020017219 | ISBN 9781138589735 (hardback) | ISBN
    9780429491443 (ebook)
Subjects: LCSH: Supercritical fluids.
Classification: LCC QC145.4.T5 P76 2021 | DDC 530.4/2--dc23
LC record available at https://lccn.loc.gov/2020017219

ISBN: 978-0-367-54935-0 (pbk)
ISBN: 978-1-138-58973-5 (hbk)
ISBN: 978-0-429-49144-3 (ebk)

DOI: 10.1201/9780429491443

Typeset in Minion Pro
by Cenveo® Publisher Services

Front cover image of Venus is Courtesy NASA/JPL-Caltech.

Contents

Preface...xi
About the Authors...xv
Useful Equations and Definitions xvii
Definitions.. xxiii

1   Some Remarks on the Gas State
    1.1   Equation of State (EOS) of Real Gases 1
          1.1.1   The Van der Waals Equation 1
          1.1.2   The Virial Equation...2
    1.2   Order in the Gas State...3
    1.3   Heat Capacity of Gases 4
          1.3.1   How Well Does This Model Work?.............................. 4
    1.4   Vibrational Raman Spectroscopy of Gases 6
    1.5   Viscosity of Gases 8
    1.6   Why Are Liquids so Difficult? 10
          1.6.1   Molecular Dynamics (MD) 10
          1.6.2   The Fundamental EOS (Section 3.3)..........................11
          1.6.3   Treat the Fluid as Gas-Like 12
          1.6.4   Treat the Fluid as Solid-Like 12
    References... 13

2   The Vapour Pressure Curve and the Liquid State Close to
    the Vapour Pressure Curve
    2.1   Classical Versus Quantum Liquids.. 15
    2.2   The Transition Across the Vapour Pressure Curve.......................... 17
    2.3   The Clausius-Clapeyron Equation 19
          2.3.1   Validity of the Clausius-Clapeyron Equation.............. 20
    2.4   The Critical Point.. 20
          2.4.1   Critical Constants and the Van Der Waals
                  Equation of State..25
    2.5   Summary... 29
    References... 30

3   Equations of State for Fluids
    3.1   Cubic EOS Based on the Van der Waals Equation 32
          3.1.1   Volume Translation of Cubic EOS... 34
    3.2   The Carnahan-Starling EOS 35
    3.3   The Fundamental EOS 36
          3.3.1   Ideal Gas Component of the Helmholtz Function........................ 36
          3.3.2   Residual Component of the Helmholtz Function 39
          3.3.3   Fitting the Helmholtz Function to the
                  Experimental Data 39
    3.4   Conclusions 41
          3.4.1   For What Fluids Is a Fundamental EOS Available? 41
          3.4.2   How Can We Test the Validity of an EOS?....................................41
          3.4.3   What Is the Best Way to Implement Your
                  Chosen EOS? 44
    References.. 46

4   The Liquid State Close to the Melting Curve (I):
**   Static Properties**
    4.1   Density and Bulk Modulus of Fluids Close to the Melting Curve.............. 47
          4.1.1   Density of Fluid Ar Close to the Melting Curve............................ 48
          4.1.2   Density and Bulk Modulus of Fluid N_2 Close to
                  the Melting Curve 49
    4.2   Elastic Neutron and X-ray Diffraction from Liquids Close to the
          Melting Curve 51
          4.2.1   Distinctions Between X-ray and Neutron
                  Diffraction Experiments 53
          4.2.2   Fourier Transform of Fluid Diffraction Data
                  to Obtain $g(r)$...55
          4.2.3   Fourier Transform of Modified Fluid Diffraction Data
                  to Obtain $g(r)$...58
          4.2.4   Comparison of Diffraction Data to Simulated Fluid
                  Structures in Reciprocal Space..61
           4.2.5   Relation Between $g(r)$, the Partition Function, Internal
                  Energy, and Pressure..63
           4.2.6   Relation Between $g(r)$ and Entropy...65
          4.2.7   Relation Between $g(r)$ and Co-ordination Number (CN)............ 66
    4.3   Short-Range Order and Phase Transitions in Fluids Close to the
          Melting Curve 67
          4.3.1   Co-ordination Number 67
          4.3.2   Liquid-Liquid Phase Transitions.. 67
    4.4   Equations to Fit the Melting Curve on the P,T Phase Diagram 69
    4.5   What Happens to the Melting Curve in the High P,T Limit?......................72
    4.6   Summary..74
    References..77

5 The Liquid State Close to the Melting Curve (II):
Dynamic Properties
5.1 Phonon Theory of Liquids 79
5.1.1 Frenkel and Maxwell Models 79
5.1.2 Prediction of Liquid Heat Capacity 82
5.2 Raman Spectroscopy of Liquids and Supercritical Fluids
Close to the Melting Curve 88
5.2.1 Grüneisen Model for Vibrational Raman
Peak Position 90
5.2.2 Hard Sphere Fluid Theory of Vibrational
Raman Peak Positions 91
5.2.3 Peak Position of Rotational Raman Spectra 93
5.2.4 Peak Intensity and Linewidth of Fluid Raman Spectra ... 93
5.2.5 Prediction of Fluid Raman Spectra Using MD 94
5.3 Brillouin Spectroscopy of Liquids Close to the Melting Curve ... 96
5.4 Inelastic Neutron and X-ray Scattering from Liquids Close to
the Melting Curve 98
5.4.1 Distinction Between Neutron and X-ray Scattering ... 98
5.4.2 The Scattered Intensity 101
5.4.3 What Can We Learn from Inelastic Neutron and X-ray
Scattering from Liquids? 103
5.5 Summary and Outlook 107
References 108

6 Beyond the Critical Point
6.1 The Widom Lines 111
6.1.1 A Simple Phenomenological Fitting Procedure for
the Widom Lines 114
6.1.2 Some Examples of Widom Line Paths 117
6.1.3 The Widom Lines as a Function of
Reduced Temperature 119
6.1.4 The Widom Lines in Relation to the Vapour
Pressure Curve 120
6.1.5 The Widom Lines as a Function of Density 121
6.2 The Fisher-Widom Line 121
6.3 The Joule-Thomson Inversion Curve 123
6.4 A General Approach to Inversion Curves 127
6.4.1 First Order Inversion Curves: Definitions 128
6.4.2 First Order Inversion Curves: Path on the P,T
Phase Diagram 132
6.4.3 Zeroth and First Order Inversion Curves: Can We
Measure Them? Do We Need to Measure Them? 134
6.4.4 Use of Zeroth Order and First Order Inversion Curves to
Verify Equations of State 136

6.5 The Frenkel Line 138
    6.5.1 Definitions of the Frenkel Line 138
    6.5.2 The Frenkel Line and the Widom Lines............................147
    6.5.3 Positive Sound Dispersion Above T_C............................148
    6.5.4 Termination of the Frenkel Line 150
6.6 Conclusions 150
References..151

7 Miscibility in the Liquid and Supercritical Fluid States
7.1 Introduction 153
7.2 Raoult's Law, Henry's Law, and the Lever Rule 154
    7.2.1 Raoult's Law and Henry's Law..154
    7.2.2 Change in Gibbs Function on Mixing of
           Raoultian Liquids...156
    7.2.3 Phase Equilibria in Miscible Fluids: The Lever Rule 158
7.3 Hildebrand Theory of Mixing 158
    7.3.1 Internal Energy of Fluid Mixtures Using
           Hildebrand Theory..158
    7.3.2 P, V, T EOS for Mixtures Using Hildebrand Theory 160
7.4 Application of the Fundamental EOS to Mixtures 162
7.5 Some Comments on Experimental Study of Supercritical Fluid
    Mixtures...163
    7.5.1 Preparation of Fluid Mixtures in the Diamond
           Anvil Cell (DAC) 163
    7.5.2 Raman Spectra of Fluid Mixtures; Cohesive
           Energy Density 164
7.6 Open Questions in the Study of Dense Fluid Mixtures 165
    7.6.1 Is Hydrophobicity an Absolute Property? 165
    7.6.2 Miscibility in the Supercritical Fluid State......................166
References..167

8 Applications of Supercritical Fluids
8.1 Applications of Supercritical Fluids in Power Generation Cycles............169
    8.1.1 Efficiency of Thermodynamic Cycles.............................169
    8.1.2 Use of Supercritical H_2O in Power Generation 170
    8.1.3 Use of Supercritical CO_2 in Power Generation...............171
    8.1.4 Use of Supercritical N_2 in Power Generation................173
8.2 Use of Supercritical Fluids in Food Processing 175
    8.2.1 Decaffeination...175
    8.2.2 Other Food Processing Applications 175
8.3 Supercritical CO_2 Cleaning and Drying 175
8.4 Chromatography...176
8.5 Crystal and Nanoparticle Growth 176
8.6 Exfoliation of Layered Materials 177
References..178

9 Supercritical Fluids in Planetary Environments
Helen E. Maynard-Casely

9.1 Introduction 181
9.2 Mineral and Material Processes with Supercritical Fluids 182
    9.2.1 Dissolution of Minerals 182
    9.2.2 Mineral Reactions 184
    9.2.3 Partition of Elements 185
9.3 Supercritical Fluids within Surface and Subsurface Environments 185
    9.3.1 Earth 186
    9.3.2 Other Terrestrial Planets 187
    9.3.3 Dwarf Planets and Icy Satellites 190
9.4 Supercritical Fluids within Planetary Interiors 190
    9.4.1 Jupiter and Saturn 191
    9.4.2 Uranus and Neptune 192
    9.4.3 Transitions in the Supercritical Fluids; Effect
          on the Gas Giants 194
9.5 Summary 194
References 194

Appendix A: Reference Data on Selected Atomic Fluids

A.1 Table of Phase Change Properties for He, Ne, and Ar 199
A.2 Phase Diagram of He 199
A.3 Phase Diagram of Ne 205
A.4 Phase Diagram of Ar 208
References 210

Appendix B: Reference Data on Selected Molecular Fluids

B.1 Table of Phase Change Properties for CH_4, CO_2,
    H_2, H_2O, and N_2 211
B.2 Phase Diagram of CH_4 211
B.3 Phase Diagram of CO_2 217
B.4 Phase Diagram of H_2 220
B.5 Phase Diagram of H_2O 226
B.6 Phase Diagram of N_2 231
References 234

Appendix C: Some Thermodynamic and Diffraction Derivations

C.1 Thermodynamic Quantities 237
    C.1.1 Application of the First Law of Thermodynamics 237
    C.1.2 Adiabatic Changes; Enthalpy 238
    C.1.3 Isothermal Changes; Helmholtz Function 238
    C.1.4 Isobaric and Isothermal Changes; Gibbs Function 239
    C.1.5 Constraints on the *P, V, T* EOS of the Ideal Fluid and the
          Condensing Fluid (Brown's Conditions) 239
C.2 Fourier Transform Treatment of Diffraction 242

Appendix D: The Diamond Anvil Cell (DAC)

D.1   Design of the DAC 245
D.2   Loading of Fluid and Fluid Mixture Samples into the DAC 247
       D.2.1   Pure Fluids 247
       D.2.2   Fluid Mixtures 249
D.3   High Temperatures in the DAC 249
       D.3.1   Resistive Heating Experiments in the DAC 250
       D.3.2   Laser Heating in the DAC 251
D.4   Pressure Measurement in the DAC 252
       References 254

Appendix E: Code for Selected Computational Problems

E.1   Boiling Transition in the van der Waals Fluid 255
       E.1.1   Estimate of P_b 256
       E.1.2   Evaluation of ΔG 257
       E.1.3   Octave Code for van der Waals' Boiling Transition 257
E.2   Prediction of Fluid Heat Capacity 260
       E.2.1   Octave Code for Heat Capacity Calculations 260

Bibliography 263

Index 267

Preface

I was motivated to write this text by my experience researching the transition(s) between liquid-like and gas-like behaviour in supercritical fluids over the past seven years and experience prior to that studying melting curves. Several points became abundantly clear to me during this period. Firstly, that there are several different research communities working on fluids who take very different approaches to the subject and do not interact very much. There is a need to reconcile the approaches taken by different research communities into one unified picture of the liquid and supercritical fluid states. Secondly, that there have been important advances in the past decade which have not been properly placed into context in a textbook alongside existing understanding of the liquid and supercritical fluid states. Thirdly, that a careful and comprehensive survey of the subject could provide some exciting new leads for future research. Some readers may discern these by reading between the lines of this text.

The textbook presented here, that has resulted from these observations, is intended to be a survey of our modern understanding of the liquid and supercritical fluid states of matter. I prioritized making a broad survey of the field rather than discussing certain minor aspects of the field in great depth; texts in the bibliography already do this.

I have only given derivations of formulae in this text if I believe that examining the derivation can significantly benefit the reader's understanding. For instance I have not given the derivation leading to the dynamic structure factor $S(Q,\omega)$ because the key information is evident from examining the result, that it is related to the time-varying real-space structure of the sample by a double Fourier transform. On the other hand, I have given the derivation for the Kechin equation as through examination of this the reader can see the extent to which it is based on first principles. I have outlined the derivation of the Lorch modification function in detail as it is from this that the reader can understand the conditions under which it will fail. In addition, I have outlined a number of experimental problems in the study of dense liquids and supercritical fluids which "everybody knows about" but which are rarely discussed in writing.

One of the principal reasons it is worthwhile to know about the properties of dense supercritical fluids is the application to planetary science. The outer layers of the major and outer planets consist principally of warm, dense, fluid mixtures confined at conditions beyond the critical point. Since I am not a planetary scientist, I am very grateful to my colleague Dr. Helen Maynard-Casely for writing a chapter (Chapter 9) devoted to the role played by supercritical fluids in planetary interiors.

Appendix E contains code for tackling certain computational problems in this field. I am also happy to provide electronic copies of the code upon request to j.e.proctor@salford.ac.uk, or to receive feedback and suggestions for improvements to future editions.

I would also like to outline here a few practical points in relation to the text. The text is written at a level suitable for researchers and postgraduate students. Therefore, certain University-level knowledge and understanding is assumed; for instance, knowledge of the distinction between ideal gases and real gases, degrees of freedom, properties of van der Waals forces, and existence of the critical point. To do otherwise would have resulted in a text of unmanageable length. Where I recap on these matters, it is to remind readers of the knowledge; I assume that they have already obtained the understanding. Some good undergraduate level texts on this subject are listed in the bibliography for the benefit of readers who need to revise.

Also in the interests of brevity, I have avoided certain topics: Glasses and the glass transition are not covered, wetting and surface tension are only covered where relevant to phase transitions and industrial applications.

The text here deals, wherever possible, with real liquids and supercritical fluids. I would like to define several matters of nomenclature. The term "fluid" will be used when making an argument that applies to liquids, gases, and supercritical fluids. The term "supercritical fluid" will be used when referring to the supercritical fluid state specifically. At conditions of supercritical temperature but subcritical density behaviour is gas-like so the fluid can in this case be referred to as a gas. We will discuss both atomic and molecular fluids in the text. In a discussion that applies to both atomic and molecular fluids the term "particle" will be used to denote 'an atom in the case of an atomic fluid and a molecule in the case of a molecular fluid.

In order to ensure that the subject matter of the text stays close to real fluids, experimental data are frequently used to illustrate key points. I am grateful to those scientists listed below who have kindly provided their data for this endeavour, and also—of course—for the NIST REFPROP database of fluid properties available online. Here, and elsewhere in the literature, what is found on NIST is referred to as experimental data. Strictly, this is not true as the information on NIST is the output from certain mathematical models (for instance, the Wagner-Setzmann fundamental equation of state) that are fitted to the experimental data using (sometimes very many) adjustable parameters. However, due to the care taken by NIST to select suitable models and only provide output in the specific P,T regions for which the models are fitted to data, I feel that it is acceptable to describe the output purely as "data" and treat it as such.

I have plotted figures in terms of pressure, unless there is a specific reason to use density instead. This is because, usually, it is pressure which is directly measured in an experiment and the density is calculated from the pressure assuming a particular equation of state. In consideration of equations of state I have chosen as the default option to plot density as a function of pressure rather than volume as a function of pressure. This is because, upon isothermal pressure increase above the critical temperature (beginning in the gas state, then passing into the supercritical fluid state followed by transition to the solid state) the density rises from close to zero to a modest finite value whilst the volume drops from close to infinity to a modest finite value. It is therefore possible to illustrate changes in density over a large P,T range much more easily than changes in volume. Densities and volumes are molar unless otherwise stated.

I would like to thank friends, colleagues, and other researchers in the field who have assisted in the preparation of this text by critically reviewing various chapters, providing advice on computer coding, photography, or rendering, or by providing their experimental or simulation data: David Beynon, Prof. Vadim Brazhkin, Cillian Cockrell, Prof. Trevor Cox, Prof. Ulrich Deiters, Prof. Martin Dove, Aidan Dunbar, Prof. David Dunstan, Dr. Jon Eggert, Dr. Valentina Giordano, Dr. Jonathan Hargreaves, Jack Holguin, Dr. Ross Howie, Michael Johnston, Dr. Stephen Lane, Dr. John Loveday, Dr. Helen Maynard-Casely, Dr. Daniel Alfonso Melendrez Armada, Prof. Andy Moorhouse, Dr. Andrew Nelms, Dr. Clemens Prescher, Dr. Ciprian Pruteanu, Liam Read, Alex Ritchie, Prof. Mario Santoro, Dr. Andrei Sapelkin, Prof. Alan Soper, Prof. Kostya Trachenko, Dr. Chenxing Wang, Prof. Lionel Wilson, and the University of Salford Maker Space.

John E. Proctor
Rawtenstall

About the Authors

John E. Proctor is a senior lecturer in physics at the University of Salford, and is head of the Materials and Physics Research Group. He specializes in condensed matter physics, particularly the study of fluids and solids under extreme pressure and temperature, principally through X-ray and neutron diffraction along with optical spectroscopy. His research is regularly published in leading international peer-reviewed journals. He completed his Ph.D. (2007) from the University of Manchester and his M.Phys. (2004) from the University of Oxford. He is one of the authors of *An Introduction to Graphene and Carbon Nanotubes* (CRC Press, 2017).

Helen E. Maynard-Casely is an instrument scientist at the Australian Nuclear Science and Technology Organisation. She specializes in planetary material research, with particular attention to the materials that make up the surfaces and interiors of the dwarf planets of our solar system. She completed her Ph.D. (2009) in high-pressure neutron scattering from the University of Edinburgh and her M.Sci. (2005) from University College London. She also works to make science accessible to all audiences, with her efforts including publication of the children's book *I Heart Pluto* (Sourcebooks Kids, 2020).

Useful Equations and Definitions

Free energies

Enthalpy: $\qquad\qquad\qquad\qquad\qquad\qquad\qquad H = U + PV$

Helmholtz function $\qquad\qquad\qquad\qquad\qquad F = U - TS = H - PV - TS$

Gibbs function $\qquad\qquad\qquad\qquad\qquad\qquad G = U - TS + PV$

Relation of internal energy to P,V,T $\qquad P + \left(\dfrac{\partial U}{\partial V}\right)_T = T\left(\dfrac{\partial P}{\partial T}\right)_V$

Derivation of fluid properties from radial distribution function $g(r)$

Energy equation $\qquad\qquad\qquad\qquad U = \dfrac{3}{2}Nk_BT + 2\pi N\rho \displaystyle\int\limits_{r=0}^{\infty} \mathcal{V}(r)g(r)r^2 dr$

Pressure equation $\qquad\qquad\qquad \dfrac{P}{\rho k_B T} = 1 - \dfrac{2\pi\rho}{3k_BT}\displaystyle\int\limits_{r=0}^{\infty}\dfrac{d\mathcal{V}}{dr}g(r)r^3 dr$

Compressibility equation $\qquad\qquad \rho k_B TK = 1 + 4\pi\rho\displaystyle\int\limits_{r=0}^{\infty}\left[g(r)-1\right]r^2 dr$

Relation between a function and it's Fourier transform

$$F(w) = \dfrac{1}{(2\pi)^{3/2}}\int f(v)e^{iv.w}d\tau$$

$$f(v) = \dfrac{1}{(2\pi)^{3/2}}\int F(w)e^{-iv.w}d\tau$$

Here, $f(v)$ is the Fourier transform of $F(w)$ and vice versa. We can of course choose how the $1/(2\pi)^{3/2}$ factor is split between the two transforms and on which exponent we place the minus sign. The changes $v.w \to wv$ and $(2\pi)^{3/2} \to \sqrt{2\pi}$ would result in the Fourier transform relationship between the scalar quantities v,w.

Clausius-Clapeyron equation

$$\frac{dP}{dT} = \frac{S_g - S_l}{V_g - V_l}$$

OR:

$$\frac{dP}{dT} = \frac{L}{T(V_g - V_l)}$$

Maxwell relations

$$\left(\frac{\partial T}{\partial V}\right)_S = -\left(\frac{\partial P}{\partial S}\right)_V$$

$$\left(\frac{\partial P}{\partial T}\right)_V = \left(\frac{\partial S}{\partial V}\right)_T$$

$$\left(\frac{\partial T}{\partial P}\right)_S = \left(\frac{\partial V}{\partial S}\right)_P$$

$$\left(\frac{\partial V}{\partial T}\right)_P = -\left(\frac{\partial S}{\partial P}\right)_T$$

Bridgman differentials

A large collection of differentials of thermodynamic properties was established by Bridgman, from which the Maxwell relations, as well as many other thermodynamic relations, can be derived. Bridgman's differentials, first published by Bridgman in *A Condensed Collection of Thermodynamics Formulas* (Harvard University Press, 1925) and subsequently reprinted in various other texts [2.8], are given below.

   P constant

$$(\partial T)_P = 1$$

$$(\partial V)_P = \left(\frac{\partial V}{\partial T}\right)_P$$

$$(\partial S)_P = \frac{C_P}{T}$$

$$(\partial Q)_P = C_P$$

$$(\partial W)_P = P\left(\frac{\partial V}{\partial T}\right)_P$$

$$(\partial H)_P = C_P$$

$$(\partial G)_P = -S$$

$$(\partial F)_P = -S - P\left(\frac{\partial V}{\partial T}\right)_P$$

From these differentials we can construct any partial derivatives we require, for instance:

$$\left(\frac{\partial V}{\partial W}\right)_P = \frac{1}{P}$$

Similarly, other parameters may be kept constant:

H constant

$$(\partial P)_H = -C_P$$

$$(\partial T)_H = V - T\left(\frac{\partial V}{\partial T}\right)_P$$

$$(\partial V)_H = -C_P\left(\frac{\partial V}{\partial P}\right)_T - T\left[\left(\frac{\partial V}{\partial T}\right)_P\right]^2 + V\left(\frac{\partial V}{\partial T}\right)_P$$

$$(\partial S)_H = \frac{VC_P}{T}$$

$$(\partial Q)_H = VC_P$$

$$(\partial W)_H = -P\left[C_P\left(\frac{\partial V}{\partial P}\right)_T + T\left[\left(\frac{\partial V}{\partial T}\right)_P\right]^2 - V\left(\frac{\partial V}{\partial T}\right)_P\right]$$

T constant

$$(\partial P)_T = -1$$

$$(\partial V)_T = -\left(\frac{\partial V}{\partial P}\right)_T$$

$$(\partial S)_T = \left(\frac{\partial V}{\partial T}\right)_P$$

$$(\partial Q)_T = T\left(\frac{\partial V}{\partial T}\right)_P$$

$$(\partial W)_T = -P\left(\frac{\partial V}{\partial P}\right)_T$$

$$(\partial U)_T = T\left(\frac{\partial V}{\partial T}\right)_P + P\left(\frac{\partial V}{\partial P}\right)_T$$

$$(\partial H)_T = -V + T\left(\frac{\partial V}{\partial T}\right)_P$$

$$(\partial G)_T = -V$$

$$(\partial F)_T = P\left(\frac{\partial V}{\partial P}\right)_T$$

G constant

$$(\partial P)_G = S$$

$$(\partial T)_G = V$$

$$(\partial V)_G = V\left(\frac{\partial V}{\partial T}\right)_P + S\left(\frac{\partial V}{\partial P}\right)_T$$

$$(\partial S)_G = \frac{1}{T}\left[VC_P - ST\left(\frac{\partial V}{\partial T}\right)_P\right]$$

$$(\partial Q)_G = VC_P - ST\left(\frac{\partial V}{\partial T}\right)_P$$

$$(\partial W)_G = P\left[V\left(\frac{\partial V}{\partial T}\right)_P + S\left(\frac{\partial V}{\partial P}\right)_T\right]$$

S constant

$$(\partial P)_S = -\frac{C_P}{T}$$

$$(\partial T)_S = \left(\frac{\partial V}{\partial T}\right)_P$$

$$(\partial V)_S = -\frac{1}{T}\left[C_P\left(\frac{\partial V}{\partial P}\right)_T + T\left[\left(\frac{\partial V}{\partial T}\right)_P\right]^2\right]$$

$$(\partial Q)_S = 0$$

$$(\partial W)_S = -\frac{P}{T}\left[C_P\left(\frac{\partial V}{\partial P}\right)_T + T\left[\left(\frac{\partial V}{\partial T}\right)_P\right]^2\right]$$

$$(\partial U)_S = -\frac{P}{T}\left[C_P\left(\frac{\partial V}{\partial P}\right)_T + T\left[\left(\frac{\partial V}{\partial T}\right)_P\right]^2\right]$$

$$(\partial H)_S = -\frac{VC_P}{T}$$

$$(\partial G)_S = -\frac{1}{T}\left[VC_P - ST\left(\frac{\partial V}{\partial T}\right)_P\right]$$

$$(\partial F)_S = \frac{1}{T}\left[PC_P\left(\frac{\partial V}{\partial P}\right)_T + PT\left[\left(\frac{\partial V}{\partial T}\right)_P\right]^2 - ST\left(\frac{\partial V}{\partial T}\right)_P\right]$$

V constant

$$\left(\partial P\right)_V = -\left(\frac{\partial V}{\partial T}\right)_P$$

$$\left(\partial T\right)_V = \left(\frac{\partial V}{\partial P}\right)_T$$

$$\left(\partial S\right)_V = \frac{1}{T}\left[C_P\left(\frac{\partial V}{\partial P}\right)_T + T\left[\left(\frac{\partial V}{\partial T}\right)_P\right]^2\right]$$

$$\left(\partial Q\right)_V = C_P\left(\frac{\partial V}{\partial P}\right)_T + T\left[\left(\frac{\partial V}{\partial T}\right)_P\right]^2$$

$$\left(\partial W\right)_V = 0$$

$$\left(\partial U\right)_V = C_P\left(\frac{\partial V}{\partial P}\right)_T + T\left[\left(\frac{\partial V}{\partial T}\right)_P\right]^2$$

$$\left(\partial H\right)_V = C_P\left(\frac{\partial V}{\partial P}\right)_T + T\left[\left(\frac{\partial V}{\partial T}\right)_P\right]^2 - V\left(\frac{\partial V}{\partial T}\right)_P$$

$$\left(\partial G\right)_V = -V\left(\frac{\partial V}{\partial T}\right)_P - S\left(\frac{\partial V}{\partial P}\right)_T$$

$$\left(\partial F\right)_V = -S\left(\frac{\partial V}{\partial P}\right)_T$$

Definitions

B_1, B_2	First and second virial coefficients
C	Compression factor
C_V, C_P	Heat capacity at constant volume, pressure per mole
c_V, c_P	Heat capacity at constant volume, pressure per particle
C.P.	Critical point
DAC	Diamond anvil cell
EOS	Equation(s) of state
EPSR	Empirical potential structure refinement
F	Helmholtz function
$f(Q)$	Atomic form factor (in diffraction)
G	Gibbs function
$g(r)$	Radial distribution function
H	Enthalpy
MD	Molecular dynamics simulation
N	Number of particles comprising the fluid sample
n	Number of moles comprising the fluid sample
N_A	Avogadro's number
P_R, V_R, T_R	Reduced (dimensionless) pressure, volume and temperature ($=1$ at the critical point, equation 2.19)
PSD	Positive sound dispersion
$\mathbf{Q}$	Scattering vector in diffraction experiment
Q	Scalar value of scattering vector, *or* heat (depending on context)
S	Entropy
$S(Q)$	Static structure factor
T.P.	Triple point
τ_R	Atomistic (Frenkel) liquid relaxation time (section 5.1.1)
U	Internal energy
$\mathcal{V}(r)$	Pair potential
VAF	Velocity autocorrelation function
Z	Fluid partition function
Z_P	Single-particle partition function
$Z(t)$	Velocity autocorrelation function

1

Some Remarks on the Gas State

In this short chapter we will discuss certain properties of ideal and real gases. The intention is not to provide a complete and self-contained treatment of the gas state (many excellent textbooks treat this subject at all levels [1.1]), but to revise certain properties that contrast with those of liquids and dense supercritical fluids. We can write down and accept a few properties of all gases right away; they have no long-range order, they do have non-zero viscosity, and they are always miscible with each other. We will now consider in a little more detail some other properties of gases.

At the end of this chapter we begin our discussion of liquids and dense supercritical fluids by outlining why it is that liquids are difficult—why we cannot understand and model their properties effectively using simple analytical methods as we do for gases or, for that matter, solids.

1.1 Equation of State (EOS) of Real Gases

The equation of state (EOS) for one mole of an ideal gas, $PV = N_A k_B T$ or $PV = RT$, is etched into the mind of every science student since they were at school. The most commonly used EOS for a non-ideal gas, the van der Waals equation (equation 1.1) [1.2] and the virial equation (equation 1.2), are also well known. Here we will briefly revise the physical principles underpinning their mathematical form.

1.1.1 The Van der Waals Equation

$$\left(P + \frac{a}{V^2}\right)(V - b) = RT \tag{1.1}$$

Equation 1.1 is based on the EOS for an ideal gas, with two corrections; those using the coefficients a and b. Let us deal with a first.

In an ideal gas it is assumed that there are no forces exerted between the gas particles. In a real gas, however, the particles constantly exert attractive van der Waals forces on each other. This has the effect of decreasing the pressure exerted by the gas on its container, compared to the case of an ideal gas; i.e., a kind of "pressure defect." In a real gas, the particles only impact the walls of the container after they have overcome the attractive van der Waals forces binding them to other particles. The pressure defect is proportional to the number of particles striking the container wall per unit time, i.e., it is proportional to the density ρ. Also, at higher density there are more particles attracting each particle that escapes to strike the container walls so in total it is proportional to ρ^2. The pressure defect is conventionally accounted for by the $\frac{a}{V^2}$ term in equation 1.1, although it would be equally valid to write $a'P^2$. The pressure defect is not sufficiently large to prevent the gas from expanding to fill the available volume of the container; all gases do this, by definition.

The second correction, b, is due to the finite volume of the gas particles. In an ideal gas the particles are assumed to have zero volume; it is a model that applies at low density. As the gas is compressed the total volume of the particles is a significant proportion of the volume V of the container. This is accounted for by writing $(V - b)$ instead of just V in equation 1.1.

Whilst other EOS for gases have been developed (for instance the Dieterici equation, Berthelot equation, and virial equation (1.2)), the van der Waals equation (1.1) performs remarkably well at modelling the equation of state of most gases over large ranges of pressure and temperature. It also successfully models the deviation in heat capacities from the ideal gas form $C_P - C_V = R$.

1.1.2 The Virial Equation

An alternative to the van der Waals equation is to implement additional terms to correct the ideal gas EOS in the form of a power series. This is the virial equation (equation 1.2). The coefficients $B_n(T)$ are called the virial coefficients. In general, they need to be temperature dependent for the EOS to have any utility in describing real gases (except for the first virial coefficient: $B_1 = 1$ always) but are not dependent on P, V. If this was permitted the virial approach would have no merit in producing a simple EOS.

$$PV = RT\left[B_1 + \frac{B_2(T)}{V} + \frac{B_3(T)}{V^2} + \cdots \right] \tag{1.2}$$

Here we have expanded in terms of volume as this is the form that we will require later in the text. An expansion in terms of pressure (terms dependent on P, P^2 etc.) is also often performed [1.3]. In the low pressure, high volume limit we can manage with only the first term (reducing to the ideal gas EOS). As we move further from the low pressure, high volume limit successively higher terms become significant in value. In this text we will make use of the second virial coefficient B_2 only.

Through rigorous statistical mechanical arguments [1.4][1.5] it can be shown that $B_2(T)$ accounts for the interactions between two particles (as opposed to three particle interactions etc.) and can be derived from the potential energy $\mathcal{V}(r)$ between a pair of particles separated by a distance r (the pair potential) as seen in equation 1.3:

$$B_2(T) = 2\pi N_A \int_0^{\infty} \left[1 - e^{-\mathcal{V}(r)/k_B T} \right] r^2 dr \tag{1.3}$$

Here, N_A is Avogadro's number.

1.2 Order in the Gas State

It is frequently stated that gases have no long-range, or short-range, order. Whilst this statement is correct, it is open to misinterpretation so deserves to be discussed in this opening chapter. Both gases and liquids exhibit no long-range order. Molecular liquids (especially those such as water with shapes and intermolecular potentials that are a long way from being spherically symmetric) often exhibit short-range order in the orientation of molecules relative to those nearby. Gases do not exhibit this. In addition, in the ideal gas limit (sufficiently low density and high temperature) there is no structure factor $S(Q)$ or radial distribution function $g(r)$ ($g(r)=1$ is expected theoretically [1.1] (Figure 4.8) and observed experimentally, for instance ref. [1.6]).

However, as the gas is compressed the volume of the gas particles does take up a significant proportion of the volume of the container (this is, after all, the justification for the $(V-b)$ term in the van der Waals EOS (equation 1.1)). In this case, a structure factor and radial distribution function for the gas do exist and the gas is no longer ideal. Figure 1.1 shows the raw neutron diffraction data for gaseous Ar at 350 K at a range of pressures demonstrating the appearance of $S(Q)$ upon pressure increase. All but the two highest pressure datapoints correspond to a subcritical density, thus demonstrating the appearance of $S(Q)$ whilst the Ar is still—unequivocally—in the gas state. The $S(Q)$

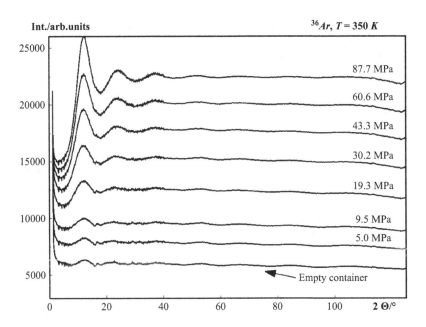

FIGURE 1.1 Raw data for the structure factor $S(Q)$ of gaseous Ar collected upon pressure increase at 350 K at subcritical density (5.0 MPa–43.4 MPa) and supercritical density (60.6 MPa and 87.7 MPa) [1.7]. Reprinted from Till Pfleiderer, Isabella Waldner, Helmut Bertagnolli, Klaus Tödheide, Barbara Kirchner, Hanspeter Huber, and Henry E. Fischer, J. Chem. Phys. **111**, 2641 (1999), with the permission of AIP publishing.

observed from dense gases correspond to dense random packing arrangements rather than arrangements with short-range order.

1.3 Heat Capacity of Gases

The heat capacity at constant volume of gases can be understood in terms of the degrees of translational, vibrational, and rotational freedom of the individual particles. In the simplest case, the monatomic gas, there are translational degrees of freedom only. This results in an internal energy per particle of $U = \frac{3}{2}k_BT$, $\frac{1}{2}k_BT$ for each degree of freedom. The heat capacity at constant volume c_V is then obtained using equation 1.4 as $\frac{3}{2}k_B$ per particle.

$$c_V = \left(\frac{\partial U}{\partial T}\right)_V \tag{1.4}$$

In a diatomic gas, there are also rotational and vibrational degrees of freedom. As a result, the heat capacity is no longer a constant as it can vary according to whether there is adequate thermal energy available to excite the quantized translational and vibrational energy levels. For a diatomic gas we expect $c_V = \frac{3}{2}k_B$ per particle at low temperature, then an increase to $\frac{5}{2}k_B$ per particle when adequate thermal energy is available to excite the two rotational degrees of freedom around the axes perpendicular to the bond axis, then an increase to $\frac{7}{2}k_B$ per particle when the vibrational degree of freedom can also be excited (if the molecule does not dissociate before that temperature). The single vibrational degree of freedom contributes k_B to the heat capacity instead of $\frac{k_B}{2}$ because at any given time the atoms possess both kinetic and potential energy.

1.3.1 How Well Does This Model Work?

Readers may have studied undergraduate textbooks in which the changes in heat capacity of gases as adequate thermal energy becomes available to excite the rotational, and then vibrational, degrees of freedom are represented in schematic diagrams as neat step functions separated by plateaux. The reality (Figure 1.2) is a bit messier, though still in agreement with the ideal gas model under the ρ, T conditions in which it would be reasonably expected to work.

Thus far, we have considered the heat capacity per particle at constant volume (c_V). However, the widest variety of experimental data available are for measurements of heat capacity at constant (1 atmosphere) pressure, expressed per mole of particles rather than per particle. Henceforth we will use C_V, C_P to denote heat capacity per mole of particles whilst c_V, c_P are per particle. For an ideal gas $c_P = c_V + k_B$. In Figure 1.2 we plot the data for C_P (available from NIST REFPROP) at 0.1 MPa for N_2, H_2, and Ar. The data are plotted from the saturated vapour temperature upwards.

Considering the simplest case, Ar, first, we observe an extremely high heat capacity upon boiling at 87 K. In the gas state close to the vapour pressure curve the heat capacity remains significantly higher than that expected for an ideal gas (5R/2). We will discuss this phenomenon in more detail in the next chapter, but for now we can note that

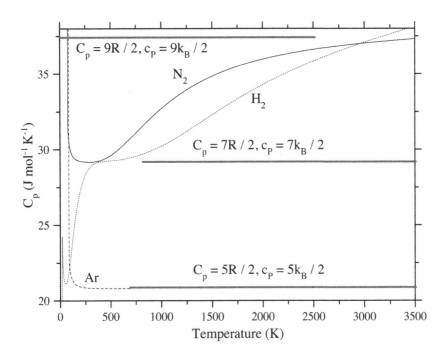

FIGURE 1.2 Heat capacities (C_P) at 0.1 MPa pressure from NIST REFPROP for N_2 (solid black line), H_2 (dotted black line) and Ar (dashed black line). The thick grey lines mark the heat capacities expected for an ideal gas with only translational freedom (5R/2), for a diatomic ideal gas with translational and rotational freedom (7R/2), and for a diatomic ideal gas with translational, rotational, and vibrational freedom (9R/2).

immediately following boiling the density of the gas may still be too high for the ideal gas model to be valid*. By 500 K the density at 0.1 MPa has decreased sufficiently for the ideal gas model to work and the heat capacity remains at the expected value of 5R/2 up until the highest temperature for which data are available.

In N_2 also, the heat capacity is extremely high upon boiling at 77 K, but as soon as boiling has taken place there is adequate thermal energy available to excite the rotational degrees of freedom resulting in C_P dropping to 7R/2 rather than 5R/2. Upon further temperature increase C_P then rises to 9R/2 as expected when the vibrational degrees of freedom become available.

In H_2, the boiling point is much lower (20 K) so immediately after boiling there is adequate thermal energy available only to excite the translational degrees of freedom and C_P drops to nearly 5R/2. Then it rises to 7R/2 when the rotational degrees of freedom can be excited; but at the highest temperatures we see it rising way beyond 9R/2 rather than remaining constant at this value. This is due to a small proportion of the H_2 molecules dissociating into atomic hydrogen, a phenomenon observed by Langmuir

---

* In a van der Waal's gas $C_P - C_V = R\left(1 + \frac{2a}{RVT}\right)$ so C_P is slightly higher than would be the case for an ideal gas [1.16]. Also, for the van der Waals gas $C_V \equiv 3R/2$, which is not obeyed for dense real gases.

as long ago as 1912 [1.8]. Upon further temperature increase the hydrogen atoms would dissociate and a plasma would be formed. These phenomena are interesting but beyond the scope of this textbook.

1.4 Vibrational Raman Spectroscopy of Gases

In all molecular fluids, and solids, the frequencies (energies) of the vibrational excited states shift as a function of pressure and temperature. Generally the shifts are small compared to the overall energy of the excited state. For instance, the Raman-active vibrational mode of H_2 (corresponding to stretching of the H-H bond) has an energy/frequency of ca. 4200 cm^{-1} (522 meV)† and shifts by about 40 cm^{-1} between ambient conditions and 50 GPa pressure as the sample goes through the fluid and solid states. However, the shifts are usually large enough to be easily resolvable with modern Raman and infrared spectrometers so can provide valuable diagnostic information about a sample.

In dense liquids, and solids, the frequencies of the vibrational excited states usually (but not always) increase upon pressure increase (at constant temperature) and decrease upon temperature increase (at constant pressure). However, in gases the opposite is true; as pressure is increased the vibrational frequencies decrease and as temperature is increased the vibrational frequencies increase. This can be understood in terms of the van der Waals forces between gas molecules. As pressure (density) increases the attractive van der Waals forces become more significant. They have the counterintuitive effect of loosening (lengthening) the intramolecular bond and hence decreasing the vibrational frequency. In the gas state the molecules are too far apart for pressure increase to have the opposite effect of compressing the intramolecular bonds and hence increasing the vibrational frequency.

Figures 1.3 and 1.4 illustrate the points above. Figure 1.3 is a Raman spectrum of the principal CH_4 vibrational mode at 300 K illustrating the accuracy in peak fitting that is achievable (ca. 0.1 cm^{-1}). Whilst the shifts in vibrational frequency discussed above may be minute compared to the absolute value of the frequency, they are easily resolvable on modern optical spectrometers. Figure 1.4 is the findings of a wide-ranging study of the change in Raman frequency of gaseous CH_4 over a wide range of conditions. The shift to lower frequency upon increasing pressure (density) is reproduced over a wide range of temperatures.

---

† It is conventional to give the frequencies (energies) measured using Raman spectroscopy using units of cm^{-1}. This is not technically correct since we are using units of cm^{-1} to describe energy, and not the wavenumber (which has to be close to zero for any excitation resulting from a first order non-resonant Raman scattering process). There is, however, logic behind this choice of units. When Raman scattering is performed experimentally, the Raman scattered photons are diffracted at a grating and the raw data recorded are the angles through which they are diffracted. This is directly determined by the relation between the wavenumber of the photon, and the line spacing on the diffraction grating. Converting into units of energy using $E = hc\tilde{v}$ requires the values of h and c. When Raman scattering was developed as an analytical tool (in the 1930s) there was still a possibility that the accepted values of these constants could change. It was therefore appropriate for scientists to report their results in units of cm^{-1} so that the values reported were independent of any assumptions about the value of h and c. Here we will use the term "Raman shift" when reporting measured frequencies (energies) using units of wavenumber.

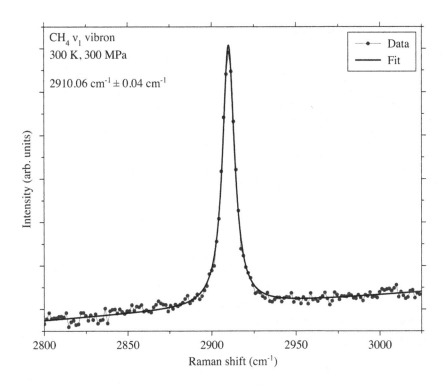

FIGURE 1.3 Raman spectrum of CH_4 vibrational mode with Lorentzian fit. Author's own data.

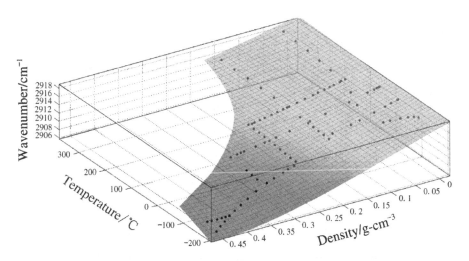

FIGURE 1.4 Raman frequency (labelled as wavenumber) of the principal Raman-active vibration in CH4 collected across a range of temperatures, and a range of subcritical densities [1.9]. Reprinted with permission from L. Shang, I.M. Chou, R.C. Burruss, R. Hu and X. Bi, J. Raman Spec. **45**, 696 (2014). Copyright © 2014 John Wiley & Sons, Ltd.

1.5 Viscosity of Gases

The viscosities of gases may be an order of magnitude lower (typically) than those of liquids, but they are not zero. In fact, their study as part of the kinetic theory of gases provides an important link between properties measurable on a macroscopic scale and the fundamental properties of the gas on a microscopic scale: The particle diameter, incorporated into the pair potential $\mathcal{V}(r)$. This can be illustrated with a simple derivation (provided, for instance in refs. [1.3][1.5]) giving the viscosity of a gas in terms of the mean thermal velocity $\langle c \rangle$. We begin with the mean free path λ of the particles composing the gas. This can be determined from the collision cross section for the particles, in turn determined from their diameter d. We will use ρ to denote the number density of particles in the gas:

$$\lambda = \frac{1}{\sqrt{2}\pi\rho d^2} \tag{1.5}$$

We also require the standard definition of viscosity (η) in terms of the frictional force between successive layers of the gas. Figure 1.5 illustrates the model we will use: A gas undergoing laminar (as opposed to turbulent) flow in the x direction but in contact with a stationary surface at $z = 0$. Particles in contact with the surface remain stationary and those at successively higher z move at higher flow velocity in the x-direction, superimposed on their thermal (Maxwellian) average velocity $\langle c \rangle$. In the notation established in

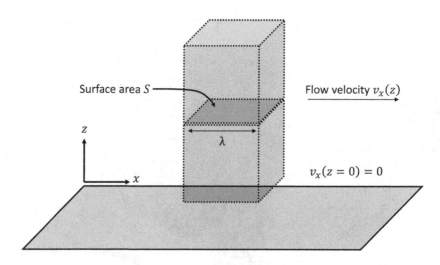

FIGURE 1.5 Schematic diagram of laminar flow of a gas. A flow velocity v_x in the x-direction is superimposed on the thermal velocities of the particles, starting from $v_x = 0$ where the gas is in contact with a stationary surface (shown) at $z = 0$ and thereafter increasing as a function of z. We consider the motion of particles across a hypothetical surface S, separating two cubic boxes with dimensions of the mean free path λ.

Figure 1.5 the frictional force resisting the relative motion of the gas layers on each side of the surface S is:

$$F = \eta S \frac{dv_x}{dz} \qquad (1.6)$$

We envisage two cubic boxes in the gas with dimensions of the mean free path λ. Per unit time, $\rho \langle c \rangle S / 4$ particles from each box will strike the surface at z_0 joining the two boxes. Those from the box above carry a total momentum x-component p_{x+} into the lower box and those from the box below carry a lower total momentum x-component p_{x-} into the upper box:

$$p_{x+} = mv_x(z_0) + m\lambda \left(\frac{dv_x}{dz} \right)_{z=z_0}$$

$$p_{x-} = mv_x(z_0) - m\lambda \left(\frac{dv_x}{dz} \right)_{z=z_0}$$

Here we have accounted for the difference in average momentum between particles in the box above and below the surface S. We are going to implicitly assume that the flow velocity in the x-direction is too small to significantly perturb the Maxwellian velocity distribution represented through $\langle c \rangle$. Since the average speed of the N_2 molecules in the air at 300 K is ca. $\langle c \rangle \approx 700$ ms^{-1}, this is usually reasonable. We therefore obtain for the net momentum transfer ΔP across the surface per unit time:

$$\Delta P = \frac{1}{4} \rho \langle c \rangle S \left[p_{x+} - p_{x-} \right]$$

$$\Delta P = \frac{1}{2} \rho \langle c \rangle S m\lambda \left(\frac{dv_x}{dz} \right)_{z=z_0}$$

Equating this to the frictional force defined in equation 1.6 we obtain:

$$\eta = \frac{1}{2} \rho \langle c \rangle m\lambda \qquad (1.7)$$

The mean velocity $\langle c \rangle$ can be derived directly from the temperature[‡] so it is thus possible to directly relate the viscosity measured on a macroscopic scale to the microscopic parameter of the particle diameter d using equation 1.5. We can also explain the—at first sight—counterintuitive experimental observations that the viscosities of gases in the low pressure limit increase as a function of temperature but are independent of pressure.

[‡] $\langle c \rangle = \sqrt{\dfrac{8 k_B T}{\pi m}}$

In addition, a more exact analysis of viscosity measurements can yield more detailed information about the pair potential than merely the particle diameter d. Assuming a spherically symmetric pair potential of the Lennard-Jones form (see later discussion following equation 2.4) we can obtain values for both the parameters in the Lennard-Jones potential ($\varepsilon, a \approx d / 2$) that allow us to model fluid behaviour also at liquid-like densities.

We can see on a fundamental level that the transport property of viscosity concerns the transfer of momentum. In an analogous way we can treat thermal conductivity as the transfer of heat and diffusion as the transfer of mass/particles across a boundary.

1.6 Why Are Liquids so Difficult?

When we consider gases and solids, we can usually understand and predict the key physical properties with reasonable accuracy beginning from first principles without resorting to computation. For instance, we can consider ballistic motion of particles in a gas and collisions with the walls of the container to obtain the EOS of an ideal gas, or we can obtain a reasonably correct phonon dispersion relation for solids by considering the atoms as balls joined by springs. In both cases these are analytical solutions that can be obtained with a pen and paper and with no recourse to computation. Why is this not possible for liquids, or for supercritical fluids at liquid-like densities?

We may formulate our description of the problem in terms of order, or in terms of potential. In a solid, the complete order simplifies matters as each atom spends all of its time near to a certain equilibrium position. In a gas the complete lack of order simplifies matters as the particles spend their time in motion, and the sample is completely isotropic on all length scales. If we need to consider the interaction between gas particles at all, it is weak enough to be treated as a perturbation.

In terms of potentials, the gas is simplified by the potential being completely constant (we may as well say zero) and the particles having only kinetic energy. The solid is simplified by the potential being constant and periodic, consisting of deep potential wells from which atoms cannot generally escape.

In liquids the potential due to the interactions between particles is not weak enough to ignore or treat as a perturbation unless we are at the lowest liquid densities in the vicinity of the critical point. On the other hand, the potential wells in which particles sit are not generally so deep that the particles cannot escape from time to time, leading to the potential changing in time and not having any long-range spatial periodicity.

Essentially, there are four different approaches to the problem, as outlined below. None of them is entirely satisfactory. This text considers separately the kind of conditions under which the different approaches are utilized.

1.6.1 Molecular Dynamics (MD)

The first approach is to accept that we cannot predict liquid properties from first principles analytically, and therefore pursue the problem computationally instead. We can run full molecular dynamics (MD) simulations of the liquid or supercritical fluid and predict properties with reasonable accuracy. For instance, one of the most difficult problems with liquids is understanding and predicting the propagation of (some) shear waves as it requires taking into account both gas-like and solid-like properties of the liquid. MD has been used to successfully model shear wave propagation in liquids for many years [1.10][1.11].

MD is still—in a way—working from first principles as the molecular dynamics simulation is working from the laws of electrostatics and classical mechanics. However, the simulation only tells us about the properties of a specific fluid under one ρ,T condition at a time. Prediction of fluid properties over a wide ρ,T range is computationally expensive.

1.6.2 The Fundamental EOS (Section 3.3)

In principle, the prediction of static and dynamic fluid properties from first principles could proceed as follows. We could write down the partition function Z_P for an individual particle in the fluid, then from this obtain the partition function Z for the fluid, then obtain the Helmholtz function $F(\rho,T)$ for the fluid from Z:

$$F(\rho,T) = -k_B T \ln Z$$

The key static and dynamic properties of the fluid can then be calculated from the Helmholtz function, for instance the pressure at a set volume/density:

$$P = -\left(\frac{\partial F}{\partial V}\right)_T$$

The problem with this approach lies at the first stage: Writing down the partition function. For a solid, a set of energy levels can be written down and the only change upon ρ,T variation is which of these energy levels are occupied. For a fluid we cannot make this simplification. Even if we could, there remains the question of how to construct the partition function for the fluid from the single particle partition function. The procedure for doing this depends on whether the particles are distinguishable. On short timescales the particles are distinguishable and on long timescales they are not. For a specific fluid the timescales for which the particles comprising the fluid are distinguishable is expected to vary by many orders of magnitude upon ρ,T change. In some cases experimental measurements have been possible of the time for which a fluid particle remains in a specific equilibrium position. For instance, this parameter in salol varies by 13 orders of magnitude upon temperature increase [1.12].

The fundamental EOS approach[§] tackles this problem by—essentially—skipping the first step. Instead of beginning with the partition function, the Helmholtz function F is obtained by fitting to the available experimental data on the static and dynamic properties which can be derived from it.

This works, but only using exceedingly complex expressions for $F(\rho,T)$ in which not just the adjustable constants, but the mathematical form of many terms in the expression

---

[§] The fundamental EOS is often referred to in the literature instead as the "reference EOS". In this text we do not use this nomenclature to avoid confusion with the "reference potential" utilized in reverse Monte Carlo / Empirical Potential Structure Refinement analysis.

are obtained empirically. Due to the empirical nature of the obtained $F(\rho,T)$, the extent to which it can be reliably used at ρ,T outside the range of the experimental data used to fit it, or to predict properties for which data was not utilized in the fitting process, is questionable.

1.6.3 Treat the Fluid as Gas-Like

Dating from the time of van der Waals, we can treat the interaction between particles in the fluid as a perturbation to gas-like behaviour. This is the approach most commonly taken. The merits of this approach in certain ρ,T regions are clear; when we are at the highest temperatures and lowest densities at which the liquid state exists the liquid has many important properties in common with a gas: Density, the ability to flow even on very short timescales, and an inability to support shear waves. All other fluids share with gases a lack of long-range order.

This approach is referred to in the literature as hydrodynamics. The various semi-empirical attempts over the decades to incorporate solid-like behaviour (elasticity) into the hydrodynamic approach [1.11][1.13] are referred to as "generalized hydrodynamics."

1.6.4 Treat the Fluid as Solid-Like

On the other hand, liquids and supercritical fluids at lower temperature and/or higher pressure have properties more similar to those of solids. The ability to flow is only realized on extremely long timescales, the density becomes close to (in the case of H_2O even exceeding) the corresponding solid phase, and shear waves can be supported. The approach beginning from similarity to solids and incorporating the gas-like properties as a correction was first put forward by Maxwell; in this text we will draw more on the atomistic approach of Frenkel. Both approaches centre on the use of some kind of structural relaxation time to describe the properties of the dense fluid. On timescales longer than this the behaviour is gas-like, on timescales shorter than this the behaviour is solid-like.

Detailed investigations to compare the (theoretical) results of both approaches are ongoing [1.14]. To conclude Chapter 1 it is appropriate to make one comment on the necessity of accounting for the high frequency solid-like (or phonon-like) excitations in the fluid. These are the excitations that contribute the most to the internal energy, as demonstrated using the equation below from ref. [1.15] which will be discussed later in Chapter 5:

$$U_L = \frac{3N(1+\gamma\beta T)}{\omega_D^3} \int_0^{\omega_D} \omega^2 \times \frac{\hbar\omega d\omega}{e^{\frac{\hbar\omega}{k_B T}} - 1}$$

Here, the contribution U_L to the fluid internal energy arising from longitudinal excitations at different frequency ω and hence energy is summed. The fluid Grüneisen parameter is γ and the thermal expansion coefficient is β. The highest frequency at which modes exist (Debye frequency) is given by ω_D. Provided that there is adequate thermal energy to excite all modes ($k_B T \gtrsim \hbar \omega_D$), the contribution to the internal energy from the highest frequency modes dominates due to the dependence of the density of states on ω^2 as well as the higher energy of each individual excitation.

References

1.1 D. Tabor, *Gases, liquid and solids*, Cambridge University Press, Cambridge (1969).

1.2 *Over de continuiteit van den gas-en vloeistoftoestand*, Ph.D. thesis of J.D. van der Waals, Universiteit Leiden, The Netherlands (1873).

1.3 P.W. Atkins, *Physical chemistry*, Oxford University Press (1982).

1.4 L.D. Landau and E.M. Lifshitz, *Statistical physics*, Pergamon Press, Oxford (1969).

1.5 W.J. Moore, *Physical chemistry*, Longman (1972).

1.6 C.J. Benmore et al., J. Phys.: Cond. Mat. **11**, 3091 (1999).

1.7 T. Pfleiderer et al., J. Chem. Phys. **111**, 2641 (1999).

1.8 I. Langmuir, *J*. Am. Chem. Soc. **34**, 860 (1912).

1.9 L. Shang, I.M. Chou, R.C. Burruss, R. Hu and X. Bi, J. Raman Spec. **45**, 696 (2014).

1.10 G. Jacucci and I.R. McDonald, Mol. Phys. **39**, 515 (1980).

1.11 J.P. Hansen and I.R. McDonald, *Theory of simple liquids*, Academic Press, Oxford (2006).

1.12 R. Casalini, K.L. Ngai and C.M. Roland, Phys. Rev. *B* **68**, 014201 (2003).

1.13 J.P. Boon and S. Yip, *Molecular hydrodynamics*, Dover Publications, New York (1991).

1.14 K. Trachenko and V.V. Brazhkin, Rep. Prog. Phys. **79**, 016502 (2016).

1.15 D. Bolmatov, V.V. Brazhkin and K. Trachenko, Sci. Rep. **2**, 421 (2012).

1.16 B.H. Flowers and E. Mendoza, *Properties of matter*, Wiley, New York (1970).

2

The Vapour Pressure Curve and the Liquid State Close to the Vapour Pressure Curve

In this chapter we will begin our discussion of the liquid state. Throughout this text, unless otherwise specified, we deal with classical liquids (and supercritical fluids). We therefore define the conditions fulfilled by a classical liquid in Section 2.1. The main part of this chapter consists of a thought experiment: Supposing that we condense a sample into the liquid state by crossing the vapour pressure curve not far below the critical point, what causes this transition and what is the minimum set of changes to properties that is required for the sample to become a liquid rather than a gas? What can we state about the vapour pressure curve from first principles? We will then conclude by looking at critical phenomena and the conditions fulfilled at the critical point.

2.1 Classical Versus Quantum Liquids

The classical liquid is defined by two conditions, which we describe here. Generally, the real fluids that we discuss in this text fulfil the classical fluid conditions (see discussion relating to He in Appendix A and H_2 in Appendix B). The first condition is that the de Broglie thermal wavelength λ of the particles comprising the liquid is small compared to the mean nearest-neighbour separation between the particles in the liquid [1.11] (equation 2.1). The temperature dependence of λ ensures that the condition is almost always met for supercritical fluids. This assumption allows us to separate the contributions to the total energy E of the system arising from kinetic (K) and potential (V) energy of the particles:

$$\lambda = \sqrt{\frac{2\pi\hbar^2}{mk_BT}}$$

$$E_i = K_j + V_k \tag{2.1}$$

In this case any kinetic energy level K_j can be combined with any potential energy level V_k to produce a different overall energy level E_i. We therefore write the partition function for the single particle, Z_P, as follows:

$$Z_P = \sum_i e^{-\frac{E_i}{k_B T}} = \sum_i e^{-\frac{K_j + V_k}{k_B T}} = \left(\sum_j e^{-\frac{K_j}{k_B T}} \right) \times \left(\sum_k e^{-\frac{V_k}{k_B T}} \right)$$

The second condition fulfilled by the classical liquid is that all particles in the liquid are in different quantum states, achieved by the number of available quantum states for the system vastly exceeding the number of particles. If this condition is met, then the equation below correctly relates the partition function Z of the system of N indistinguishable particles to the single-particle partition function Z_P [2.1]:

$$Z = \frac{1}{N!}(Z_P)^N = \frac{1}{N!}\left(\sum_j e^{-\frac{K_j}{k_B T}} \right)^N \times \left(\sum_k e^{-\frac{V_k}{k_B T}} \right)^N \tag{2.2}$$

In this case the Helmholtz function F for the system, derived in the usual manner ($F = -k_B T \ln Z$), is:

$$F = -N k_B T \left[\ln \left(\sum_j e^{-\frac{K_j}{k_B T}} \right) + \ln \left(\sum_k e^{-\frac{V_k}{k_B T}} \right) \right] + k_B T \ln(N!)$$

Using Stirling's formula* to deal with the final term, we obtain:

$$F = -N k_B T \left[\ln \left(\sum_j e^{-\frac{K_j}{k_B T}} \right) + \ln \left(\sum_k e^{-\frac{V_k}{k_B T}} \right) \right] + N k_B T \ln(N) - N k_B T$$

All terms in the summations above corresponding to quantum states which are accessible (i.e., adequate thermal energy is available, $k_B T \gtrsim K_j$, $k_B T \gtrsim V_k$) are roughly equal to unity so, when the number of accessible quantum states vastly exceeds the number of particles the last two terms above are insignificant and we obtain:

$$F = -N k_B T \left[\ln \left(\sum_j e^{-\frac{K_j}{k_B T}} \right) + \ln \left(\sum_k e^{-\frac{V_k}{k_B T}} \right) \right] \tag{2.3}$$

The Helmholtz function has thus also been separated additively into contributions from kinetic and potential energy of the particles as a result of the classical approximation.

---

* $\ln(N!) = N \ln N - N$

2.2 The Transition Across the Vapour Pressure Curve

Let us consider a condensation transition across the vapour pressure curve not too far below the critical point (say $T \approx 0.8T_C$). Upon pressure increase or temperature decrease, the transition from the gas state to the liquid state takes place at the P,T point where the Gibbs functions of the gas and liquid states have equal value. At this point there is a (usually significant) increase in density, from which changes in other properties follow. We should begin by understanding the density increase in terms of the pair potential between the particles comprising the fluid. This function gives the potential energy V between two particles in the fluid as a function of their separation r. For the simplest fluid, that composed of spherically symmetric particles interacting only via van der Waals forces, the pair potential has the exact form given in equation 2.4 and Figure 2.1 (the Lennard-Jones potential). Parameters in equation 2.4 are a, the diameter of the fluid particles, and ε, the depth of the potential well.

$$V(r) = 4\varepsilon \left[\left(\frac{a}{r} \right)^{12} - \left(\frac{a}{r} \right)^{6} \right] \qquad (2.4)$$

The only real systems for which the Lennard-Jones potential is exactly correct are noble gas fluids, but Lennard-Jones potentials are also used as approximations for the intra- and intermolecular bonding in fluids composed of covalently bonded molecules such as

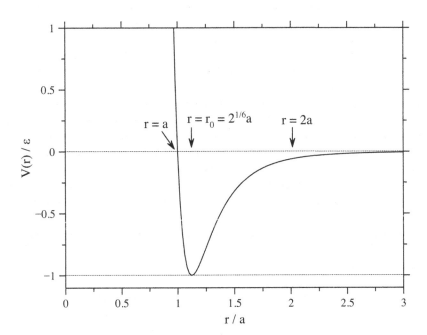

FIGURE 2.1 The Lennard-Jones potential (equation 2.4) shown as a plot of reduced potential $V(r)/\varepsilon$ against reduced separation r/a.

fluid N_2. In fluids where particles are charged the attractive part must have a significantly longer range, and in fluids composed of particles that are not spherically symmetric the potential must depend on the relative orientation of the particles as well as their separation. Nonetheless, for all fluids that condense into the liquid state the pair potential must fulfil the conditions listed in equations 2.5–2.7 and therefore have a mathematical form similar to that shown in Figure 2.1 for the Lennard-Jones potential.

$$\lim_{r \to \infty} \mathcal{V}(r) = 0 \qquad\qquad (2.5)$$

$$\lim_{r \to 0} \mathcal{V}(r) = +\infty \qquad\qquad (2.6)$$

$$\frac{d\mathcal{V}}{dr} = 0 \text{ at } r = r_0, \text{ where:}$$

$$\mathcal{V}(r_0) = -\varepsilon \qquad\qquad (2.7)$$

Equation 2.5 is necessary for $\mathcal{V}(r)$ to reflect the reality that particles do not interact at all in the $r \to \infty$ limit and equation 2.6 to prevent atomic overlap and the violation of the Pauli exclusion principle. Equation 2.7 specifies the fact that, for condensation into the liquid state to take place the potential must be attractive at some finite separation r_0. This attractive potential is the result of different interactions in different fluids: van der Waals forces, hydrogen bonding, electrostatic attraction between particles with unlike charges, etc.

The step increase in density upon condensation results from particles sitting in the potential well due to the attractive part of the pair potential shown in Figure 2.1. The lowest density a fluid can have whilst remaining in the liquid state is the density at the critical point. For a Lennard-Jones fluid this density corresponds to $r \approx 2a$ (as labelled on Figure 2.1) so for all other points in the liquid state the particle separation lies further into the potential well. A number of other parameters undergo discontinuous changes upon condensation. Principal parameters are summarized in Table 2.1.

Importantly, there are also many properties that do not change, or undergo limited changes, upon condensation. In particular, both liquids and dense gases can support longitudinal sound waves (though the speed of sound does significantly increase upon condensation). However, the liquid state in this part of the phase diagram, just like the

TABLE 2.1   Parameters changing when the vapour pressure curve is crossed at $T \approx 0.8T_C$

Quantity	Change Upon Condensation
Density ρ	Discontinuous increase
Gibbs function G	Discontinuous change in the gradient $\frac{dG}{dP}$ or $\frac{dG}{dT}$
Entropy S	Discontinuous decrease
Heat capacities at constant pressure and volume C_P and C_V	Rise on approaching the condensation point from gas state, then step increase upon condensation.

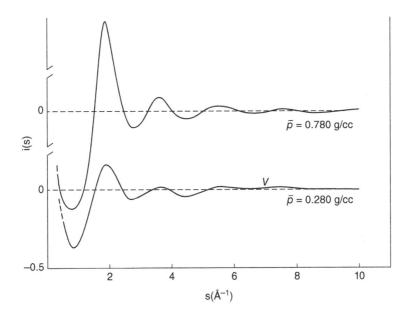

FIGURE 2.2 The intensity function $i(S)$ (directly derived from the structure factor) of gaseous (lower) and liquid (upper) Ar, close to the vapour pressure curve at 148 K [2.3]. Reprinted from P.G. Mikolaj and C.J. Pings, J. Chem. Phys. **46**, 1401 (1967), with the permission of AIP publishing.

gas state, cannot support shear waves. The structure factor $S(Q)$ and radial distribution function $g(r)$ of the fluid do change upon condensation but retain broadly the same form. Figure 2.2 shows the structure factor of Ar at 148 K ($T_C = 150.7$ K), immediately before and after the condensation transition. Condensation causes a significant increase in intensity of the first peak but otherwise little change to the mathematical form of the function, compared to the change in a gas–crystalline solid or liquid–crystalline solid transition. In was even proposed in 1993 that the condensation transition at temperatures subcritical but close to the critical point should be considered as a gas–gas phase transition [2.2].

2.3 The Clausius-Clapeyron Equation

The reductions in entropy and volume that always take place upon condensation from the gas state into the liquid state allow conclusions to be drawn about the path of the vapour pressure curve on the P,T phase diagram from first principles, in the form of the Clausius-Clapeyron equation. The vapour pressure curve marks the line where the Gibbs function of the liquid (l) and gas (g) states are equal: $G_l = G_g$. Therefore, if we alter P,T by moving a short distance along the vapour pressure curve then we must be able to write $dG_l = dG_g$ for this small shift in P,T. Hence,

$$\left(\frac{\partial G_l}{\partial P}\right)_T dP + \left(\frac{\partial G_l}{\partial T}\right)_P dT = \left(\frac{\partial G_g}{\partial P}\right)_T dP + \left(\frac{\partial G_g}{\partial T}\right)_P dT \qquad (2.8)$$

We can use the standard thermodynamic relations

$$\left(\frac{\partial G}{\partial P}\right)_T = V \text{ and } \left(\frac{\partial G}{\partial T}\right)_P = -S$$

To obtain the Clausius-Clapeyron equation for the gradient of the vapour pressure curve:

$$\frac{dP}{dT} = \frac{S_g - S_l}{V_g - V_l} \tag{2.9}$$

This equation is derived from very basic physical principles of the conservation of energy, second law of thermodynamics, definition of the Gibbs function (Appendix C) with no additional assumptions or approximations. Therefore, we can draw immediate and reliable conclusions about the vapour pressure curve; since the gas state always has higher entropy and higher volume than the liquid state, the vapour pressure curve (unlike the melting curve) always has a positive gradient on the P,T phase diagram.

2.3.1 Validity of the Clausius-Clapeyron Equation

The Clausius-Clapeyron equation can also be written in terms of the latent heat of vaporization L. Since $L = T(S_g - S_l)$, we can obtain [1.3]:

$$\frac{dP}{dT} = \frac{L}{T(V_g - V_l)} \tag{2.10}$$

All the quantities in this equation have been measured accurately for a range of fluids so the efficacy of the Clausius-Clapeyron equation has been extensively tested.

In addition, we may note that we could apply the Clausius-Clapeyron equation to a solid-liquid phase transition, a solid-gas phase transition, or for that matter, a first order phase transition between two solid states. In Chapter 4 we will utilize it to derive the Kechin equation for the P,T path of melting curves. However, we may only apply it to first order phase transitions. Whilst we did not assume a first order transition in the derivation, in a second order transition the entropy and volume of the two phases are both equal when the transition is made; therefore the Clausius-Clapeyron equation would give an undefined value for the gradient of a second order phase transition line.

2.4 The Critical Point

Earlier we discussed the condensation to the liquid state in terms of the quantum well formed as a result of the attractive component of $V(r)$ (equation 2.4). This begs the question—what happens when $k_B T \gg \varepsilon$? In this case there can be no condensation/boiling transition or higher order transition resembling it. Fundamentally, the critical temperature T_C of a fluid marks the point where the following (approximate) relation holds [1.1][1.16][2.2][2.4]:

$$k_B T_C \approx \varepsilon \qquad (2.11)$$

As the critical point is approached from lower temperature the latent heat and volume change of the transition gradually decline to zero, and above the critical point the Widom lines exist for a finite distance on the P,T phase diagram (Section 6.1). As a consequence of the aforementioned trends as the critical point is approached, the surface energy of a liquid droplet in equilibrium with the vapour in the section of the V,P phase diagram in which the liquid and vapour phases co-exist (shown later in Figure 2.6) also tends to zero. The size of the liquid droplets in the co-existence region diverges, with the droplets reaching diameters which are enormous on an atomic scale. The phenomenon of critical opalescence is caused by the size of these droplets becoming comparable to the wavelength of visible light.

We would like to understand why the condition defined in equation 2.11 marks a "tipping point" at which there is such a fundamental change in fluid properties from those of a condensing fluid to those of a fluid exhibiting only a single state. To tackle this problem properly one would have to begin by writing down a partition function for the fluid that is mathematically correct and complete over a wide ρ,T range. As discussed in Section 1.6.2, this is not possible. However, there are systems displaying analogous critical temperature phenomena that have mathematically tractable partition functions, allowing us to gain an understanding which is extremely useful. These are systems of magnetic moments which are interacting and/or are in an externally applied magnetic field.

A very simple example displaying the critical temperature phenomenon is an individual ferromagnet (magnetic moment μ) in an external magnetic field B, with only two possible orientations—parallel and antiparallel to the field. The energy levels available are thus $\pm\mu B$ and the partition function is thus:

$$Z = e^{+\frac{\mu B}{k_B T}} + e^{-\frac{\mu B}{k_B T}} = 2\cosh\left(\frac{\mu B}{k_B T}\right)$$

Now consider an array of N non-interacting ferromagnets, of which N_1 are in the $+\mu B$ level and N_2 are in the $-\mu B$ level. The total magnetization M is obtained as follows:

$$N_1 = \frac{Ne^{+\frac{\mu B}{k_B T}}}{2\cosh\left(\frac{\mu B}{k_B T}\right)}$$

$$N_2 = \frac{Ne^{-\frac{\mu B}{k_B T}}}{2\cosh\left(\frac{\mu B}{k_B T}\right)}$$

$$M = \mu N_1 - \mu N_2 = \mu N \tanh\left(\frac{\mu B}{k_B T}\right) \qquad (2.12)$$

Thus if we plot M as a function of temperature at constant B, it stays almost exactly constant at $M = \mu N$ at low temperature before declining sharply towards zero commencing at a certain critical temperature, corresponding to $x = k_B T / \mu B = 1 / e$. Figure 2.3 plots the function $y = \tanh(1/x)$ to illustrate this point.

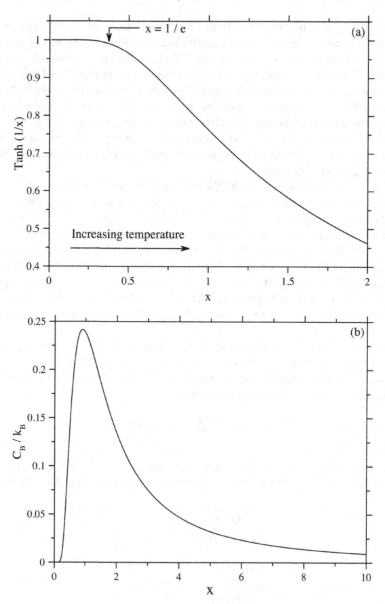

FIGURE 2.3   (a) Plot of $y = \tanh(1/x)$ illustrating change in behaviour at $x = 1/e$ corresponding to a critical temperature. (b) Heat capacity of isolated ferromagnet(s) at constant B as a function of $x = k_B T / \mu B$.

At this point the heat capacity in a constant magnetic field C_B displays a maximum. This can be obtained from the partition function via standard thermodynamic relations[†]:

$$\frac{C_B}{k_B} = \frac{1}{x^2}\left[1 - \tanh\left(\frac{1}{x}\right)\right]$$

Where $x = k_B T / \mu B$ as defined earlier. Figure 2.3 (b) plots C_B / k_B as a function of x, i.e., increasing temperature, showing the spike at the critical temperature.

A stronger analogy to the vapour pressure curve critical point may be obtained from the case of interacting magnets. This problem has been solved (but only in two dimensions) by Onsager [2.5][2.6], obtaining a critical temperature and a singularity in the heat at the critical temperature. Thus, even if we cannot perform an exact theoretical calculation to obtain the fluid behaviour at the termination of the vapour pressure curve, we see that our observations are what we would expect from comparison to simpler systems that are solvable theoretically. As Ashcroft and Mermin [2.7] noted, the liquid-vapour critical point compared to magnetism "present[s] quite strong analogies and give[s] rise to quite similar theoretical difficulties."

The changes in various properties that take place in the vicinity of the vapour pressure curve at significantly subcritical temperature become fundamentally different when the critical temperature is approached—see, for instance, the trends in the measured heat capacity shown in Figure 2.4 for a model spherically symmetric fluid composed of neutral particles (Ar). At significantly subcritical temperature it is essentially a step function, becoming a spike at the critical temperature. At significantly subcritical temperature the change in heat capacity is gradual and monotonic until the moment the vapour pressure curve is reached; in the vicinity of the critical temperature the heat capacity changes drastically throughout a finite region around the vapour pressure curve. We would like to understand these phenomena.

At the end of Chapter 1 we outlined why modelling and understanding liquids is difficult and the different approaches to this problem. To understand the critical phenomena, we will take the approach of treating the fluid as gas-like. We can demonstrate how these effects can be understood in general form using the van der Waals equation (equation 1.1). This has the advantages of being mathematically simple, and that both the parameters in the EOS can be related directly back to the fluid properties—finite particle size and attraction between particles. We note, however, that a variety of other approaches would lead to a P,V,T EOS with the same general features explaining the critical phenomena. We could employ any one of the myriad of cubic EOS based on the van der Waals equation that have been developed over the decades, or a fundamental EOS for a real fluid, or for that matter we would just work directly with the experimental data.

[†] The average thermal energy $\langle E \rangle$ is obtained via: $\langle E \rangle = k_B T^2 \frac{\partial \ln Z}{\partial T}$ then the heat capacity $C = \frac{\partial E}{\partial T}$

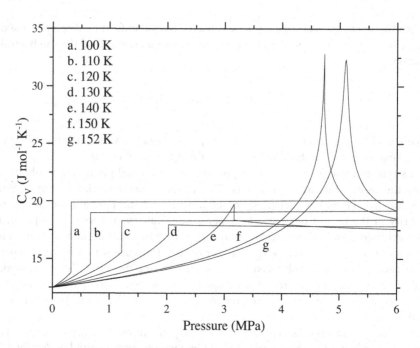

FIGURE 2.4 Trends in heat capacity C_V of fluid Ar as the vapour pressure curve is crossed upon isothermal pressure increase at 100 K, 110 K, 120 K, 130 K, 140 K, and 150 K. In addition, a transition just into the supercritical regime at 152 K is shown (the critical temperature is 150.9 K). Data have not been offset or rescaled and are obtained from NIST REFPROP.

In Figure 2.5 we show, at constant pressures, plots of reduced volume V_R calculated using the van der Waals EOS as a function of reduced temperature T_R as the vapour pressure curve is crossed along different isobars. At significantly subcritical conditions we observe a step increase in the volume upon boiling but otherwise no narrow changes in the vicinity of the vapour pressure curve. This is what we observe when we boil the kettle. In the vicinity of the critical point at $P_R = 0.98$, however, we observe significant changes in volume as the vapour pressure curve is approached from either side.

The trends in many other parameters can be derived from the P, V, T EOS, for instance the heat capacity (shown below) [1.4].

$$C_P - C_V = -T \frac{\left(\frac{\partial P}{\partial T} \right)_V^2}{\left(\frac{\partial P}{\partial V} \right)_T} \qquad (2.13)$$

As the critical point is approached $\left(\frac{\partial P}{\partial V} \right)_T$ becomes very small in the vicinity of the vapour pressure curve; that is, the sample becomes extremely compressible (we can see this from the shape of the isotherms in Figure 2.6). As we shall see later, the derivative is zero at the critical point. Hence there is a large spike in C_P in this area. Similar arguments can be made regarding the other parameters that undergo drastic changes in the vicinity of the critical point.

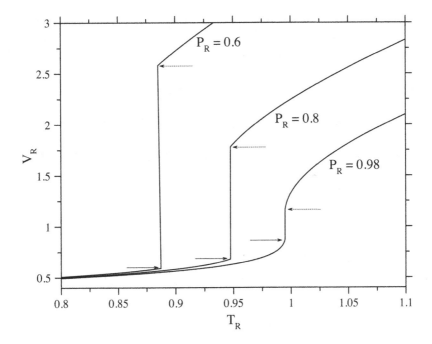

FIGURE 2.5 Reduced volume V_R plotted as a function of reduced temperature T_R as the vapour presure curve is crossed along several subcritical isobars. The reduced volumes in the saturated liquid and saturated vapour states are indicated by the solid and dotted arrows respectively. The van der Waals loops produced by the EOS have been deleted following calculation of the vapour pressure curve using the program given in Appendix E.

2.4.1 Critical Constants and the Van Der Waals Equation of State

We introduced the van der Waals equation of state (EOS) at the beginning of Chapter 1; to produce this EOS two modifications are made to the ideal gas EOS. One correction accounts for the finite non-zero volume of the gas particles, and one accounts for the attractive van der Waals forces between particles. When isotherms are plotted out on the V,P diagram (Figure 2.6) we see that the existence of the critical point and key parameters are obtained naturally from the van der Waals EOS. The isotherms plotted are obtained by rearranging equation 1.1 to consider the $P(V)$ relation for one mole of fluid:

$$P = \frac{RT}{V-b} - \frac{a}{V^2} \tag{2.14}$$

One important observation from Figure 2.6 is that the van der Waals EOS predicts what is always observed experimentally: that the critical point marks the lowest density that

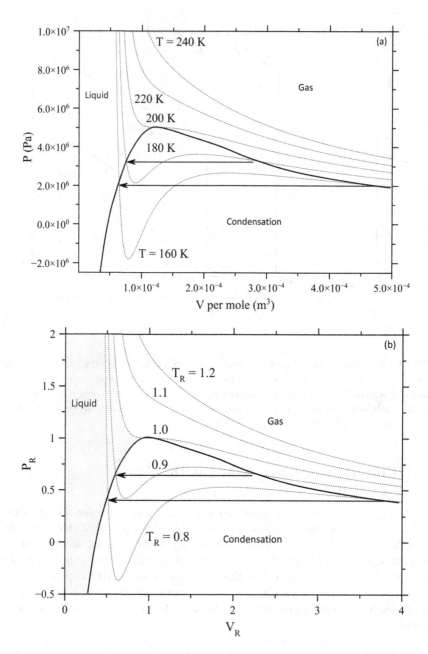

FIGURE 2.6 (a) Isotherms plotted out using the van der Waals EOS (equation 2.6) on the V,P phase diagram for a substance with $T_C = 200$ K and $P_C = 5.0$ MPa. (b) The same isotherms plotted on the reduced phase diagram. The condensation transition takes place roughly following the arrows.

a sample can have whilst remaining in the liquid state. The conditions that are only satisfied at the critical point are clearly the following:

$$\left(\frac{\partial P}{\partial V}\right)_T = 0 \tag{2.15}$$

$$\left(\frac{\partial^2 P}{\partial V^2}\right)_T = 0$$

The fulfilment of these conditions at the critical point only is also what is always observed experimentally; it is not just a mathematical peculiarity of the van der Waals EOS. Applying these conditions to equation 2.14 we obtain:

$$b = V_C / 3 \tag{2.16}$$

$$a = \frac{9}{8} V_C R T_C \tag{2.17}$$

Substituting into equation 2.14 we obtain:

$$P_C = \frac{3}{8} \frac{R T_C}{V_C} \tag{2.18}$$

If we fix any two of the critical parameters (P_c, V_c, T_c) then the set of equations 2.16–2.18 results in the values of the remaining parameter, as well as the values of a, b, being fixed. The critical compression factor ($P_C V_C / R T_C$) is fixed at $3/8$. We should note that the van der Waals equation does not generally predict the remaining critical parameter in agreement with experimental data, which shows that the critical compression factor does vary somewhat. For instance, in the case of Ar the measured values are: $T_C = 150.7$ K, $V_C = 0.0746$ Lmol⁻¹. According to equation 2.18 this would result in $P_C = 18.9$ MPa when in fact it is just 4.9 MPa. Many other cubic EOS based on the van der Waals equation have been developed to rectify this problem (reviewed in Chapter 3).

It is common to express pressure, volume, and temperature in terms of their values relative to those at the critical point. These parameters are referred to as the reduced pressure P_R, reduced volume V_R, and reduced temperature T_R (equation 2.19). In this case the van der Waals EOS simplifies to the form given in equation 2.20. Figure 2.6 (b) shows the plot in terms of the reduced parameters.

$$P_R = P / P_C \tag{2.19}$$

$$V_R = V / V_C$$

$$T_R = T / T_C$$

$$P_R = \frac{8 T_R}{3 V_R - 1} - \frac{3}{V_R^2} \tag{2.20}$$

We would like to take a closer look at the transition across the vapour pressure curve as treated with the van der Waals EOS. Considering the regions marked "liquid" and "gas" in Figure 2.6 to begin with; in these regions the van der Waals equation predicts physically realistic EOS at both subcritical and supercritical temperatures in the liquid and gas states. However, in the region marked "condensation" there is a region where the bulk modulus of the fluid is predicted to be negative; an increase in pressure causes an increase in volume. This feature of the van der Waals EOS (and other cubic EOS) is referred to as a van der Waals loop and clearly is not physically correct.

Instead, we may understand the phenomenon of boiling or condensation as follows. The black curves on Figure 2.6 mark the V,T points where the Gibbs functions of the liquid and gas states are equal, which can be expressed in terms of an integral of the P,V,T EOS across the transition, from a point A in the liquid state to a point B in the gas state (equation 2.21 [1.5][2.8]). So as far as the thermodynamics of the bulk sample are concerned, the boiling/condensation transition should occur along the lines marked with the arrows and defined by this integral, which is known as the Maxwell construction. If we (for instance) cause condensation by increasing density in an isothermal experiment the system would follow the van der Waals isotherm in the gas phase, then follow the arrow whilst condensation takes place and more of the sample turns into the liquid state at constant pressure upon further volume reduction.

$$\Delta G = \int_A^B VdP = 0 \qquad (2.21)$$

However, in actual fact there is an energetic cost to inducing the boiling/condensation transition due to the surface energy of the droplets of liquid/gas formed in the transition. The transition may therefore take place under different pressures at a given temperature depending on the direction of the transition—within the limits set by the requirement not to enter the P,T regions where the van der Waals EOS predicts a negative bulk modulus. The spinodal curve is the curve joining the maxima of the van der Waals loops denoting the limit of the region where the gas state is metastable, then joining the minima of the van der Waals loops denoting the limit of the region where the liquid state is metastable.

This matter is outlined in Figure 2.7, where we consider the boiling and condensation transitions along the $T_R = 0.9$ isotherm from Figure 2.6. In an isothermal transition at $T_R = 0.9$ transition a. corresponds to the highest pressure that could be required for condensation to occur upon pressure increase (i.e., when the sample has gone right to the end of the metastable gas isotherm). Condensation transitions can take place at pressures between transition a. and transition b. Transition b. corresponds to the pressure at which the Gibbs functions of the liquid and gas states are equal. The region between transitions b. and c. corresponds to pressures at which the boiling phase transition would take place, with c. marking the lowest pressure at which boiling could take place upon isothermal pressure decrease.

The pressure at which the Gibbs functions of the liquid and gas states are equal on the van der Waals isotherm such as that in Figure 2.7 (transition b.) is one that cannot be

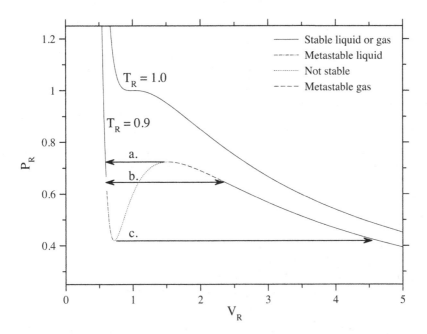

FIGURE 2.7 Van der Waals isotherms at $T_R = 0.9$ and $T_R = 1.0$. In the $T_R = 0.9$ isotherm the regions are marked where the sample is stable in the liquid state, metastable in the liquid state, metastable in the gas state, and stable in the gas state. The central part of the isotherm (dotted line) corresponds to a state which does not exist.

solved algebraically but can easily be solved computationally for the van der Waals EOS as outlined in Appendix E. Computational solution is also possible for all cubic EOS exhibiting van der Waals loops that are reviewed in Chapter 3, as long as care is taken to identify the correct roots when the equation is solved for V.

2.5 Summary

In this chapter we have outlined the conditions a classical liquid must fulfil. Most importantly, a classical liquid is one in which we can separate the contributions to the total energy of the system that arise from kinetic and potential energy of the particles comprising the liquid. In this case, it is also possible to separate the Helmholtz function of the liquid into contributions arising from the kinetic energy and potential energy of the particles. The temperature dependence of the de Broglie thermal wavelength of the particles ensures that supercritical fluids always fulfil the classical criteria.

We have introduced the concept of the pair potential $\mathcal{V}(r)$ using the simple example of the Lennard-Jones potential. The Lennard-Jones potential is exactly correct only for van der Waals-bonded systems of spherically symmetric particles, but all condensing fluids must have a pair potential that shares the key features of the Lennard-Jones potential of being zero at large r, attractive close to some equilibrium value of r, and strongly repulsive in the $r \rightarrow 0$ limit.

From the nature of $\mathcal{V}(r)$ the first order condensation phase transition of the vapour pressure curve follows, as does the fact that this curve ends at a critical point. We have explored what properties change, and what properties do not change, upon boiling/condensation. As the critical point is approached from lower P,T, the changes in properties when the curve is crossed become less significant, relative to the changes in properties as the curve is approached.

The gradient of the vapour pressure curve on the P,T phase diagram can be obtained from first principles (the Clausius-Clapeyron equation), and the path of the curve on the P,T phase diagram can be obtained computationally from the various semi-empirical cubic EOS developed to model the liquid and gas states. Code for this computation is given in Appendix E.

References

2.1 F. Mandl, *Statistical physics*, second edition, Wiley (1988).

2.2 S.M. Stishov, JETP Letters **57**, 189 (1993).

2.3 P.G. Mikolaj and C.J. Pings, J. Chem. Phys. **46**, 1401 (1967).

2.4 C. Vega, L.F. Rull and S. Lago, Phys. Rev. E **51**, 3146 (1995).

2.5 L. Onsager, Phys. Rev. **65**, 117 (1944).

2.6 T.D. Schultz, D.C. Mattis and E.H. Lieb, Rev. Mod. Phys. **36**, 856, (1964).

2.7 N.W. Ashcroft and N.D. Mermin, *Solid state physics*, Harcourt (1976).

2.8 F.W. Sears, *An introduction to thermodynamics, the kinetic theory of gases, and statistical mechanics*, Addison-Wesley, London (1959).

3

Equations of State for Fluids

Consider a liquid following (for instance) an isothermal P,T path from the melting curve to the vapour pressure curve; the changes in properties as this P,T path is followed are drastic. Figure 3.1 shows, as an example, the most basic static property: Density. On isothermal pressure increase at subcritical temperature we see roughly a doubling of the liquid density between condensation and the melting curve. If we include the supercritical fluid region as well the changes are even greater. The inset to Figure 3.1 shows the density of supercritical fluid Ar upon isothermal pressure increase from ambient pressure to the melting curve at $1.2T_C$, rising from the low pressure limit of zero in the gas state to about 40 mol L^{-1}.

The development of equations of state (EOS) that can accurately model fluid properties in a wide P,T range is necessary for applications such as the simulation of industrial processes and planetary interiors. This is an ongoing challenge, due to the fundamental difficulties presented by the liquid and supercritical fluids states which we reviewed in Section 1.6.

We can begin by reviewing the various approaches developed so far to model the EOS of fluids. We will consider the trade-offs involved in the methodology to produce an EOS and which of the different EOS proposed over the years work well for different fluids under different P,T conditions.

We should begin by understanding the reasons for the much greater complexity of fluid EOS compared to their solid counterparts. A fluid EOS should work at supercritical temperature in the vicinity of the critical point (where the compressibility is almost infinite as the Widom line is crossed) and in the vicinity of the melting curve (where the compressibility is virtually zero). Similar statements can be made for most other fluid properties and/or their derivatives. In addition, whilst liquids are not always miscible, liquid solutions form far more frequently than solid solutions. Understanding and modelling solutions/mixtures (introduced in Chapter 7) is really key to the use of liquids and supercritical fluids in industry (Chapter 8) and to understanding their behaviour in planetary interiors (Chapter 9). So it is often desirable (but hard) to produce EOS that can be applied to fluid mixtures (Sections 7.3.2 and 7.4).

We shall review the different approaches beginning with the simplest (cubic EOS based on the van der Waals equation), then the Carnahan-Starling EOS, and then finally the fundamental EOS approach—the most complex, in which the equation has over 50

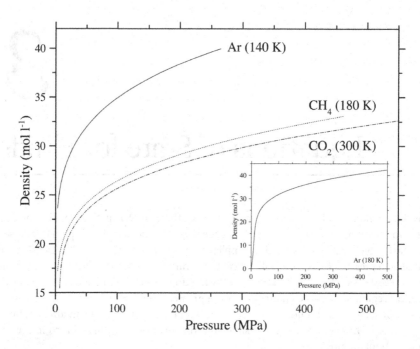

FIGURE 3.1   Density of liquid Ar ($T_C = 150.7$ K), CH$_4$ ($T_C = 190.6$ K) and CO$_2$ ($T_C = 304.1$ K) upon isothermal pressure increase from the condensation point to the highest pressure for which high accuracy data are available from NIST REFPROP. Inset: Density of supercritical fluid Ar from ambient pressure upwards at 180 K ($1.2T_C$).

adjustable parameters. We shall focus on the trade-offs between simplicity, accuracy, relation to first principles (and hence reliability of extrapolation), and invertibility.[*]

3.1 Cubic EOS Based on the van der Waals Equation

We discussed the van der Waals EOS and virial EOS in Chapters 1 and 2. However, the application of these EOS to liquids and dense supercritical fluids is limited. The virial EOS is useful to study the behaviour of fluids in the high volume limit, but it would require the inclusion of a large number of terms to have any wider utility. The van der Waals EOS is useful to study liquids in a qualitative manner in the vicinity of the critical point, but it is fundamentally unsuited to the study of liquids in general. This is because of its inability to reproduce the experimentally observed critical compression factors of real fluids (we saw in Chapter 2 that the critical compression factor obtained using the van der Waals EOS is fixed at 3 / 8).

---

[*] Invertibility is the ability to rearrange the equation into an analytical expression for any desired variable. For instance, the van der Waals EOS is given in the form $P(V,T)$ in equation 3.1 but can be rearranged into the form $T(V,P)$. For many other cubic EOS with the form given in equation 3.2 this is not possible so a computational process would be required to calculate T from V,P. Even the van der Waals EOS is not invertible to obtain V: Solution of an equation cubic in V is required.

Over the decades, various attempts have been made to produce cubic EOS retaining the $(V-b)$ term representing repulsion between particles in the van der Waals EOS (reproduced below as equation 3.1) but replacing the a/V^2 term (attraction between particles) with a temperature-dependent term.

We will review here a fraction of the different cubic EOS proposed over the decades. A complete treatment of all the different cubic EOS and their efficacy would be beyond the scope of this text. The reader is referred to other reviews for this [3.1][3.2][3.3][3.4]. It is, however, possible to make some useful general comments about cubic EOS. These EOS have the form given in equation 3.2 where a',b,c,d are constants. These constants, and the mathematical form of the function $\alpha(T)$, are chosen empirically. In all cases the volume V is per mole.

$$P = \frac{RT}{V-b} - \frac{a}{V^2} \tag{3.1}$$

$$P = \frac{RT}{V-b} - \frac{a'\alpha(T)}{V^2+cV+d} \tag{3.2}$$

Two general comments should be made at this stage. Firstly, the van der Waals EOS and all other EOS taking the form of equations 3.1 or 3.2 fail at low volume ($V \leq b$). If we reduce the volume sufficiently, the $(V-b)$ term in the van der Waals EOS becomes negative and the equation yields meaningless results [3.5] (a negative pressure required to reduce the volume further, see Figure 3.2).

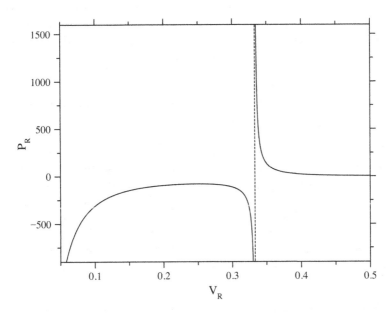

FIGURE 3.2 Reduced pressure P_R (solid line) obtained from the van der Waals EOS plotted as a function of reduced volume V_R (equation 2.20), at the critical temperature ($T_R = 1$). The singularity at $V_R = 1/3$ is indicated with a dashed line.

TABLE 3.1 Cubic EOS taking the mathematical form of equation 3.2. $a, b, a', m, K, \Omega_a, X$ are constants calculated for different fluids using the methodology outlined in the citations.

Name	Mathematical Form	$\alpha(T)$	Citation
van der Waals	$P = \dfrac{RT}{V-b} - \dfrac{a}{V^2}$	1	[1.2] (1873)
Redlich-Kwong	$P = \dfrac{RT}{V-b} - \dfrac{a'\alpha(T)}{V-b}$	$\dfrac{1}{\sqrt{T}}$	[3.6] (1949)
Redlich-Kwong-Soave	$P = \dfrac{RT}{V-b} - \dfrac{a'\alpha(T)}{V(V-b)}$	$\left[1 + m\left(1 - \sqrt{\dfrac{T}{T_C}}\right)\right]^2$	[3.7] (1972)
Peng-Robinson	$P = \dfrac{RT}{V-b} - \dfrac{a'\alpha(T)}{V(V+b)+b(V-b)}$	$\left[1 + K\left(1 - \sqrt{\dfrac{T}{T_C}}\right)\right]^2$	[3.8] (1976)
Patel-Teja	$P = \dfrac{RT}{V-b} - \dfrac{\alpha(T)}{V(V+b)+c(V-b)}$	$\Omega_a \times \dfrac{R^2 T_c^2}{P_c} \times \left[1 + X\left(1 - \sqrt{\dfrac{T}{T_c}}\right)\right]^2$	[3.9] (1982)

Secondly, inversion of these EOS to obtain volume as a function of pressure $V(P)$ must be done with caution. When $T < T_C$ we may calculate, for a given pressure, the volume of the saturated liquid and the volume of the saturated vapour. This is done by finding the roots of the cubic equation for $V(P)$ which is generally done computationally. However, it is only for the van der Waals EOS that one can reliably state that the smallest of the three roots obtained is the saturated liquid volume, the next is a mathematical artefact, and the largest is the saturated vapour volume. For other EOS with the form given in equation 3.2 the roots do not necessarily fall in this order. Table 3.1 lists the principal cubic EOS developed over the decades with the mathematical form given in equation 3.2.

Beginning with the Redlich-Kwong equation, one clear advantage is the mathematical simplicity. The price paid for this is a lack of accuracy, particularly in the vicinity of the critical point. Just like the van der Waals equation, the critical compression factor is fixed. For the Redlich-Kwong equation it is $1/3$. This is better than the van der Waals equation but still significantly higher than what is observed experimentally for most fluids. The Redlich-Kwong-Soave, Peng-Robinson, and Patel-Teja EOS, in increasing order of complexity, give more scope to fit closely to data due to the larger number of adjustable constants. We refer to the van der Waals EOS as a two-constant EOS and the Redlich-Kwong, Redlich-Kwong-Soave, Peng-Robinson, and Patel-Teja EOS as three-constant EOS (referring to the adjustable constants in the final term accounting for attractive interactions between particles).

3.1.1 Volume Translation of Cubic EOS

All of the EOS listed in Table 3.1 can be modified via changing the volume by a constant $(V \rightarrow V \pm \Delta V)$. This procedure has the effect of shifting the isotherms on the

P,V phase diagram such as Figure 2.6 to the left or right. So a systematic error in representation of sample volumes (particularly for smaller volumes in the liquid state) can be corrected without changing the location of the vapour pressure curve on the P,T phase diagram if this is already correct. See, for example, application to the Peng-Robinson EOS in ref. [3.10].

3.2 The Carnahan-Starling EOS

We saw earlier (equation 1.3) how the second virial coefficient can be calculated from the pair potential $\mathcal{V}(r)$. In principle one could, for a given pair potential, calculate successive virial coefficients to produce an ever-more-accurate EOS. Unfortunately, for any realistic pair potential the problem swiftly becomes intractably computationally complex. The exception is the hard sphere system [3.11]. For this case, it makes sense to define the virial expansion from equation 1.2 in terms of the packing fraction η instead of directly in terms of the volume. In this case, since $\eta = NV_S / V$ (V_S is the volume of an individual sphere) we obtain an equation equivalent to equation 1.2:

$$PV = RT\left[B_1 + B_2'\eta + B_3'\eta^2 ...\right]$$

Where $B_1 = 1$ but is included to clarify notation. Note that we have dropped the temperature dependence of the virial coefficients in equation 1.2. The contribution of Carnahan and Starling [3.12] was to take the first six virial coefficients that had been calculated to date (for the hard sphere system), round them to their nearest integer values and observe that these numbers followed a simple geometric progression. The virial coefficient values utilized in Carnahan and Starling's paper were:

$$PV = RT\left[1 + 4\eta + 10\eta^2 + 18.36\eta^3 + 28.2\eta^4 + 39.5\eta^5\right]$$

The following geometric series represents these values after rounding to the nearest integer:

$$PV = RT\left[1 + \sum_{n=1}^{\infty}\left(n^2 + 3n\right)\eta^n\right]$$

This entire series (summed up to infinity, not just up to $n = 5$) can be exactly evaluated algebraically to give:

$$PV = RT\frac{1 + \eta + \eta^2 - \eta^3}{\left(1-\eta\right)^3}$$

This equation is the Carnahan-Starling EOS. Note that this expression does still reduce to $PV = RT$ in the low density limited as required since in this case $\eta \to 0$. As given, this equation is only suitable as an analytical model for the P,V,T EOS of computer simulations of hard sphere systems. To apply the equation to real systems (for instance in ref. [3.13]) η is given a simple temperature-dependent form and an additional temperature-dependent term, with both fitted to EOS data with a small number of adjustable parameters:

$$PV = RT\left[\frac{1+\eta(T)+\eta(T)^2-\eta(T)^3}{(1-\eta(T))^3}+\frac{4\eta(T)\tau(T)}{T}\right]$$

3.3 The Fundamental EOS

An entirely different approach which has been developed following the ideas of Wagner and co-workers since 1989 [3.14][3.15][3.16] is to construct an equation giving the Helmholtz function F for the fluid in terms of density ρ and temperature T, the "fundamental equation[†]." In theory this could be derived from the partition function Z for the fluid ($F(\rho,T)=-k_BT\ln Z$). In reality we cannot obtain the partition function so must construct an equation for $F(\rho,T)$ in a semi-empirical manner.

The Helmholtz function is not experimentally measurable, but expressions to predict many parameters that are measurable (including the P,V,T EOS) can be derived from the fundamental equation using standard thermodynamic relations, as shown later in Table 3.2. The fundamental equation for F consists of two components, one (F_I) represents the Helmholtz function of the sample as an ideal gas at the specified ρ,T conditions. The other, the residual part F_R, represents the deviation from ideal gas behaviour (equation 3.3). As shown in Chapter 2 (equation 2.3), this division is a natural consequence of the separation of the total energy of the system into contributions arising from kinetic and potential energy of the particles comprising the system. The contribution described here, and elsewhere in the literature, as the ideal gas component is that arising from kinetic energy of the particles—in an ideal gas, the only component—and the residual component describes the contribution arising from potential energy.

$$F(\rho,T)= F_I(\rho,T)+F_R(\rho,T) \tag{3.3}$$

We deal firstly with the ideal gas component, $F_I(\rho,T)$. We will see that even this part of the fundamental equation cannot be constructed entirely from first principles.

3.3.1 Ideal Gas Component of the Helmholtz Function

Using classical thermodynamics, the ideal gas component $F_I(\rho,T)$ of the Helmholtz function can be related to the enthalpy $H_I(T)$ and the entropy $S_I(\rho,T)$ of the ideal gas (equation 3.4):

$$F_I(\rho,T)= H_I(T)-PV-TS_I(\rho,T) \tag{3.4}$$

[†] The fundamental EOS is often referred to in the literature instead as the "reference EOS". In this text we do not use this nomenclature to avoid confusion with the "reference potential" utilized in reverse Monte Carlo/Empirical Potential Structure Refinement analysis.

For an ideal gas, enthalpy is independent of density and depends only on temperature. We express the enthalpy and entropy relative to their values at some reference conditions T_0, ρ_0 (usually ambient conditions) (equation 3.5).

$$H_I(T) = H_I(T_0) + \Delta H_I(T) \tag{3.5}$$

$$S_I(T) = S_I(\rho_0, T_0) + \Delta S_I(T)$$

Dealing first with the enthalpy, we may relate this to the isobaric heat capacity C_P for a reversible process (this is appropriate since we are dealing with an ideal gas):

$$\Delta H_I(T) = \int_{T_0}^{T} C_P \, dT$$

The entropy is dealt with using a standard identity for the entropy of an ideal gas[‡] to obtain:

$$\Delta S_I(\rho, T) = \int_{T_0}^{T} \frac{C_V dT}{T} - R \int_{\rho_0}^{\rho} \frac{d\rho}{\rho}$$

Putting all this together and setting the (arbitrary) origin of $H_I(T_0) = 0$ and $S_I(\rho_0, T_0) = 0$ we obtain:

$$F_I(\rho, T) = \int_{T_0}^{T} C_P \, dT - RT + RT \int_{\rho_0}^{\rho} \frac{d\rho}{\rho} - T \int_{T_0}^{T} \frac{(C_P - R)dT}{T} \tag{3.6}$$

In Chapter 1 we discussed experimental data for the heat capacities and saw that the variation of heat capacity as a function of temperature does not—in reality—consist of neat step functions. We therefore use an empirical expression for C_P obtained to fit to experimental data. We may, for example, use equation 3.7 (ref. [3.16][3.17] for CH_4). The constants n_i, θ_i are adjusted to fit the selected experimental data. Thus we proceed by substituting for C_P.

$$C_P = Rn_0 + R \sum_{i=1}^{5} \frac{n_i \left(\dfrac{\theta_i}{T}\right)^2 e^{\theta_i/T}}{\left(e^{\theta_i/T} - 1\right)^2} \tag{3.7}$$

The resulting integrals in equation 3.6 succumb to the substitution $x = e^{\theta_i/T} - 1$ and a standard integral from the literature[§]:

$$\int_{T_0}^{T} C_P \, dT = R(T - T_0)n_0 + R \sum_{i=1}^{5} n_i \left[\frac{\theta_i}{e^{\theta_i/T} - 1} - \frac{\theta_i}{e^{\theta_i/T_0} - 1} \right]$$

[‡] $dS = C_V \dfrac{dT}{T} + R \dfrac{dV}{V}$, see for instance ref. [2.8].

[§] $\displaystyle \int \frac{\ln x \, dx}{(a+bx)^2} = -\frac{\ln x}{b(a+bx)} + \frac{1}{ab} \ln\left(\frac{x}{a+bx}\right)$

$$\int_{T_0}^{T} \frac{(C_P - R)dT}{T} = R(n_0 - 1)\ln\left(\frac{T}{T_0}\right)$$

$$+ R\sum_{i=1}^{5} n_i \left[\frac{\theta_i}{T(e^{\theta_i/T} - 1)} + \frac{\theta_i}{T} - \frac{\theta_i}{T_0(e^{\theta_i/T_0} - 1)} - \frac{\theta_i}{T_0} + \ln\left(\frac{e^{\theta_i/T_0} - 1}{e^{\theta_i/T} - 1}\right)\right]$$

Putting this all together we obtain:

$$F_I(\rho, T) = R(T - T_0)n_0 + RT\left[\ln\left(\frac{\rho}{\rho_0}\right) - 1\right] + RT(1 - n_0)\ln\left(\frac{T}{T_0}\right)$$

$$+ R\sum_{i=1}^{5} n_i \left[\frac{\theta_i}{1 - e^{-\theta_i/T_0}}\left(\frac{T}{T_0} - 1\right) + T\ln\left(\frac{e^{\theta_i/T} - 1}{e^{\theta_i/T_0} - 1}\right)\right] \tag{3.8}$$

To conclude our treatment of the ideal gas part of the Helmholtz function, we need to switch to the reduced ideal gas Helmholtz function $F_I'(\rho', \mu)$ defined in equation 3.9. The residual part of the Helmholtz function and total Helmholtz function are similarly defined in reduced form. The reduced density ρ' and inverse reduced temperature μ are defined relative to the critical density ρ_C and temperature T_C (*not* ρ_0, T_0). The constants θ_i are also reduced.

$$F_I'(\rho', \mu) = \frac{F_I(\rho, T)}{RT} \tag{3.9}$$

$$F_R'(\rho', \mu) = \frac{F_R(\rho, T)}{RT}$$

$$F'(\rho', \mu) = F_I'(\rho', \mu) + F_R'(\rho', \mu)$$

$$\rho' = \frac{\rho}{\rho_c}$$

$$\mu = \frac{T_C}{T}$$

$$\theta_i' = \frac{\theta_i}{T_C}$$

The equation for $F_I'(\rho', \mu)$ can be written in the form:

$$F_I'(\rho', \mu) = \ln(\rho') + a_1 + a_2\mu + a_3\ln(\mu) + \sum_{i=1}^{5} n_i \ln\left[1 - e^{-\theta_i'\mu}\right] \tag{3.10}$$

Where, in this case, the coefficients a_{1-3} are given by:

$$a_1 = n_0 + \ln\left(\frac{\rho_C}{\rho_0}\right) + (1 - n_0)\ln\left(\frac{T_C}{T_0}\right) + \sum_{i=1}^{5} n_i \left[\frac{\theta_i' T_C}{T_0}\left(\frac{1}{1 - e^{-\theta_i' T_C/T_0}}\right) - \ln\left(e^{\theta_i' T_C/T_0} - 1\right)\right]$$

$$a_2 = -\frac{T_0 n_0}{T_C} - \sum_{i=1}^{5} n_i \theta_i' \left[\frac{1}{1 - e^{-\theta_i T_C / T_0}} + 1 \right]$$

$$a_3 = n_0 - 1 \tag{3.11}$$

We decided earlier to set the enthalpy (H) and entropy (S) at reference conditions T_0, P_0 to zero. On the other hand, the enthalpy and entropy are both parameters that can be calculated from the reduced Helmholtz function (see Section 3.3.3) so we need to ensure that the reduced Helmholtz function we have obtained is self-consistent. Therefore, the constants a_1, a_2 are not calculated using equation 3.11 but are instead adjusted to ensure that the ideal gas components of the enthalpy (H_I) and entropy (S_I) predicted by the reduced Helmholtz function are zero at T_0, P_0.

3.3.2 Residual Component of the Helmholtz Function

The reduced residual component $F_R'(\rho, T)$ cannot be expressed in a mathematical form obtained from first principles. Instead, a form such as that shown in equation 3.12 is chosen. It has been chosen to have terms depending monotonically on ρ', μ as well as 2-dimensional Gaussian components which are included in order for the fundamental equation to be able to reproduce the sharp peaks in parameters existing in the vicinity of the critical point. The large number of terms included in equation 3.12 for $F_R'(\rho, T)$ is referred to as the "bank of terms."

$$F_R'(\rho', \mu) = \sum_{i=1}^{13} (\rho')^{d_i} x_i \mu^{t_i} + \sum_{i=14}^{36} (\rho')^{d_i} x_i \mu^{t_i} e^{-\left[(\rho')^{c_i} \right]} + \sum_{i=37}^{40} (\rho')^{d_i} x_i \mu^{t_i} e^{-g_i} \tag{3.12}$$

The terms g_i give the two-dimensional Gaussian lineshapes:

$$g_i = \alpha_i (\rho' - \Delta_i)^2 + \beta_i (\mu - \gamma_i)^2$$

The coefficients d_i, c_i, α_i are set to certain integer values, t_i set to integer or half-integer values, $\Delta_i = 1$ and γ_i take values close to 1. The terms x_i can take any real value during the fitting process described in the next section.

3.3.3 Fitting the Helmholtz Function to the Experimental Data

A number of experimentally observable properties can be obtained from the reduced Helmholtz function $F'(\rho', \mu)$ using standard thermodynamic relations. A representative selection are shown in Table 3.2 and the expressions for further properties are given in refs. [3.16] and [3.18]. In the expressions shown, the contributions to the observed property arising from the ideal gas part and residual part of the reduced Helmholtz function are separated.

We described already how the parameters are obtained for the ideal gas part of the Helmholtz function (equations 3.10 and 3.11). The large number of adjustable parameters x_{1-40} in the residual part are obtained by least squares fitting of the expressions derived

TABLE 3.2 Expressions for various properties in terms of the components of the reduced Helmholtz function [3.16]. Adapted from U. Setzmann and W. Wagner, J. Phys. Chem. Ref. Data **20**, 1061 (1991), with the permission of AIP Publishing, with additional data from ref. [3.18].

Property	Relation to $F'(\rho',\mu)$
Pressure P (or: compression factor)	$\dfrac{P}{\rho RT} = 1 + \rho'\left(\dfrac{\partial F_R'}{\partial \rho'}\right)_\mu$
Internal energy U	$\dfrac{U}{RT} = \mu\left[\left(\dfrac{\partial F_I'}{\partial \mu}\right)_{\rho'} + \left(\dfrac{\partial F_R'}{\partial \mu}\right)_{\rho'}\right]$
Enthalpy H	$\dfrac{H}{RT} = 1 + \mu\left[\left(\dfrac{\partial F_I'}{\partial \mu}\right)_{\rho'} + \left(\dfrac{\partial F_R'}{\partial \mu}\right)_{\rho'}\right] + \rho'\left(\dfrac{\partial F_R'}{\partial \rho'}\right)_\mu$
Entropy S	$\dfrac{S}{R} = \mu\left[\left(\dfrac{\partial F_I'}{\partial \mu}\right)_{\rho'} + \left(\dfrac{\partial F_R'}{\partial \mu}\right)_{\rho'}\right] - F_I'(\rho',\mu) - F_R'(\rho',\mu)$
Gibbs function G	$\dfrac{G}{RT} = 1 + \rho'\left(\dfrac{\partial F_R'}{\partial \rho'}\right)_\mu + F_I'(\rho',\mu) + F_R'(\rho',\mu)$
Isochoric heat capacity C_V	$\dfrac{C_V}{R} = -\mu^2\left[\left(\dfrac{\partial^2 F_I'}{\partial \mu^2}\right)_{\rho'} + \left(\dfrac{\partial^2 F_R'}{\partial \mu^2}\right)_{\rho'}\right]$
Isobaric heat capacity C_P	$\dfrac{C_P}{R} = -\mu^2\left[\left(\dfrac{\partial^2 F_I'}{\partial \mu^2}\right)_{\rho'} + \left(\dfrac{\partial^2 F_R'}{\partial \mu^2}\right)_{\rho'}\right] + \dfrac{\left[1 + \rho'\left(\dfrac{\partial F_R'}{\partial \rho'}\right)_\mu + \rho'\mu\left(\dfrac{\partial^2 F_R'}{\partial \rho'\partial \mu}\right)\right]^2}{1 + 2\rho'\left(\dfrac{\partial F_R'}{\partial \mu}\right)_{\rho'} + \rho'^2\left(\dfrac{\partial^2 F_R'}{\partial \rho'^2}\right)_\mu}$
Speed of sound v	$\dfrac{v}{RT} = 1 + \rho'\left(\dfrac{\partial F_R'}{\partial \rho'}\right)_\mu + \rho'^2\left(\dfrac{\partial^2 F_R'}{\partial \rho'^2}\right)_\mu - \dfrac{\left[1 + \rho'\left(\dfrac{\partial F_R'}{\partial \rho'}\right)_\mu + \rho'\mu\left(\dfrac{\partial^2 F_R'}{\partial \rho'\partial \mu}\right)\right]^2}{\mu^2\left[\left(\dfrac{\partial^2 F_I'}{\partial \mu^2}\right)_{\rho'} + \left(\dfrac{\partial^2 F_R'}{\partial \mu^2}\right)_{\rho'}\right]}$
Second virial coefficient	$B_2 = \dfrac{1}{\rho_C}\lim_{\rho\to 0}\left(\dfrac{\partial F_R'}{\partial \rho'}\right)_\mu$
Third virial coefficient	$B_3 = \dfrac{1}{\rho_C^2}\lim_{\rho\to 0}\left(\dfrac{\partial^2 F_R'}{\partial \rho'^2}\right)_\mu$

from the Helmholtz function (examples given in Table 3.2) to the available experimental data. In addition, the residual Helmholtz function (equation 3.12) is constrained to ensure that the expression for the pressure (in Table 3.2) behaves appropriately at the critical point:

1. When $\rho = \rho_C$ and $T = T_C$, the expression must agree very closely with the experimentally determined critical pressure. A key advantage of the fundamental EOS

over the van der Waals and other cubic EOS is that much closer agreement is possible with the experimentally measured critical parameters.

2. Ensuring that equation 2.15 is satisfied at the critical point.

The description given here is a crude simplification of the fitting process; a full description would be beyond the scope of this textbook, so readers are referred to specialist journal articles to obtain this [3.14][3.16].

3.4 Conclusions

In this final section, we will answer (or at least attempt to answer) questions about fluid EOS using the content covered in this chapter.

3.4.1 For What Fluids Is a Fundamental EOS Available?

We have seen that the fundamental EOS does generally give the best results, i.e., the results that most closely match the experimental data. The fundamental EOS method may involve a large empirical element, but it is no worse than the other methods in this regard. If an EOS is required, the fundamental EOS should therefore be used unless there is a specific reason to do otherwise, such as a need for a simple analytic expression for the P,V,T EOS to avoid computation. On the other hand, it must be self-evident from the description in the previous section that constructing a fundamental EOS that is fit for purpose is a substantial research project requiring a large amount of experimental data and computation. As a result, fundamental EOS covering a large ρ,T range are only available for a select group of fluids.

Fundamental EOS are available for most simple fluids, natural gas components, and refrigerants, though in some cases the P,T regime covered by the fundamental EOS is limited. In addition to EOS for pure fluids, some work has been done to produce a fundamental EOS simultaneously fitted to data on, and applicable to, certain groups of similar fluids; For instance, non-polar fluids [3.19], polar fluids [3.20], and light alkanes [3.21]. Fundamental EOS have even been produced for the Lennard-Jones fluid by fitting the empirical parameters to simulation "data" ([3.22] and refs. therein).

3.4.2 How Can We Test the Validity of an EOS?

Once we have an EOS for a given substance, how can we evaluate whether or not it is fit for purpose? Ideally, we would choose the EOS that is based most closely on first principles. Unfortunately, as we have seen in this chapter, due to the complexity of the fluid state there are no EOS available and closely based on first principles that work well over a wide ρ,T range. Failing that, we can evaluate an EOS by testing it against experimental data. The fundamental EOS are generally tested against large quantities of experimental data within the specified ρ,T range—arguably they make other semi-empirical EOS redundant. However, the cubic EOS do have the advantage of mathematical simplicity, so it is worth examining how they compare to the fundamental EOS.

Below we conduct some limited testing of different cubic EOS against the fundamental EOS for Ar and CH_4 by comparing the vapour pressure curve and the $\rho(P)$ relationship along two different isotherms $(0.8T_C, 1.2T_C)$. We can compare the output from a selection of EOS listed earlier (Peng-Robinson, Patel-Teja, and Redlich-Kwong-Soave) to the output from the fundamental EOS (Section 3.3).

Figure 3.3 shows the termination of the vapour pressure curves for Ar and CH_4 predicted using the various EOS on the P,T phase diagram. The curves predicted by the various EOS are in close agreement all the way down to the triple point; as we can see the only differences between the curves drawn using different EOS are a small variation in the predicted critical point and errors of ca. 0.05 MPa/0.2 K in the vapour pressure curve position. In Figure 3.4 we can see the ability of the different cubic EOS to predict the $\rho(P)$ relationship along the example isotherms chosen. Looking at these figures we can see that, at least as conventionally formulated, the cubic EOS are better at predicting the path of the vapour pressure curve on the P,T phase diagram than at getting the densities correct. Furthermore, it is not possible to make a definitive statement that some EOS are better than others. All work well under certain conditions.

Directly testing against experimental data may be preferable, but it is not possible if experimental data are not available for the ρ,T region of interest or if the data have large errors or are not reliable. Figure 3.5 shows an extreme case of this problem: C_2H_6 (ethane) at 300 K. A fundamental EOS has been produced for C_2H_6 [3.23], yet it is only adequately backed by experimental data up to 70 MPa (the circles are selected points from the fundamental EOS to illustrate the region in which it is backed by the experimental data). Ethane remains in the liquid state up to 2500 MPa at 300 K. Various EOS agree closely with the fundamental EOS and experimental data up to 70 MPa then rapidly diverge from each other upon further pressure increase.

Clearly, from our physical intuition we can conclude that there are certain conditions that a fluid EOS must meet. It must simplify to $PV = RT$ in the low density limit, and the high density limit cannot give $V \to 0$ as this would violate the Pauli exclusion principle. In 1960 Brown worked through these intuitive conditions and presented them in a systematic way [3.27]. Since his original paper is hard to obtain, Brown's conditions are derived in Appendix C of this text. It has been pointed out, on the other hand, that these conditions could in some cases fail if the fluid exhibits a density anomaly [3.28] (for instance the negative thermal expansion of H_2O between 273 K and 277 K at ambient pressure).

Furthermore, there are characteristic curves along which certain properties take the same value as an ideal gas under these conditions; one zeroth order curve (the Zeno curve) and three first order curves (the Boyle curve, Joule-Thomson curve, and Amagat curve) as far as P,V,T properties are concerned. These curves exist for all condensing fluids but are often terminated in practice by intersecting with the melting curve (see phase diagrams in Appendices A and B). A fluid EOS should therefore be able to predict a physically reasonable P,T path for these curves. Since these curves begin in the subcritical region but extend to extremely high pressure (often $\sim 50P_C$) and temperature (often $\sim 2.5T_C$) they are covered in detail in Chapter 6 (Sections 6.3 and 6.4), in which a unified approach is taken to their description and their use to verify EOS is discussed in some more detail.

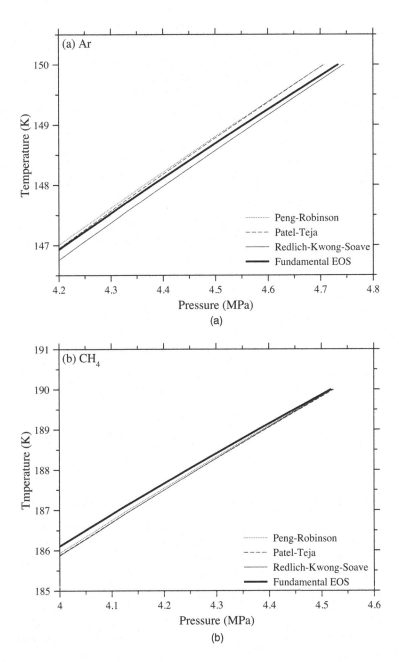

FIGURE 3.3 Termination of vapour pressure curves on the P,T phase diagram predicted by various EOS for (a) Ar and (b) CH_4.

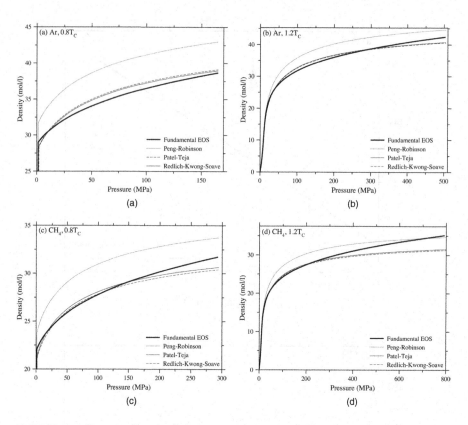

FIGURE 3.4 P, ρ EOS plotted out along isotherms for Ar at $0.8T_C$ (a) and $1.2T_C$ (b) and for CH_4 at $0.8T_C$ (c) and $1.2T_C$ (d).

3.4.3 What Is the Best Way to Implement Your Chosen EOS?

Once you have decided which EOS is best to use for your application, what is the best way to implement it? The simple 2- and 3-parameter cubic EOS discussed at the beginning of this chapter can be implemented with minimal programming, even for tasks such as finding the vapour pressure transition by integrating over the van der Waals loops (Appendix E). The computational complexity of fundamental EOS, however, is considerable. Publications describing the fundamental EOS and listing the parameters for a single simple fluid are often over 50 pages in length [3.16][3.29] and have not been digitized. It is therefore best to avoid reinventing the wheel by obtaining existing data and/or code.

If only a small number of data are required, for instance the vapour pressure curve for a pure substance or the P, V EOS at a small number of isotherms, then the free online resources described in the bibliography are likely to be adequate. At the time of writing the online access to NIST REFPROP [3.30] and to the ThermoC code [3.25] are particularly useful, giving access to the fundamental EOS for a wide variety of simple fluids

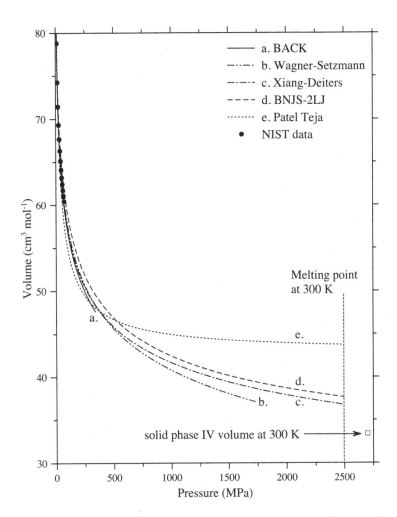

FIGURE 3.5 P,V EOS of C_2H_6 at 300 K [3.24]. Black circles are selected data points from the fundamental (Wagner-Setzmann) EOS available from NIST REFPROP for pressures at which it is backed by experimental data. Line a. is the Boublik-Alder-Chen-Kreglewski EOS, b. is the extrapolation of the fundamental EOS, c. is the Xiang-Deiters EOS, d. is the Boublík–Nezbeda + Jacobsen–Stewart EOS, and e. is the Patel-Teja EOS. Lines a-e are calculated using the ThermoC software [3.25]. The vertical dashed line marks the melting point at 300 K and the open square marks the volume of the solid state upon crystallization at 300 K [3.26]. Adapted from ref. [3.24] with permission.

(Ar, CH_4, etc.) and industrially important fluids (e.g., refrigerants). The ThermoC online interface also gives access to fundamental EOS data for mixtures.

If a large amount of data are required then there exist various other options. The ThermoC code is, at the time of writing, available free of charge to academic institutions and the REFPROP code to run offline is available for a modest charge. Other software available for a charge includes FLUIDCAL (First GmbH/Ruhr Universität Bochum [3.31]) and HYSYS (Aspen [3.32]). The HYSYS software includes EOS calculations as part of a wider package of chemical process simulation software. Additionally, a large amount of data are searchable online at the Dortmund Databank [3.33].

References

3.1 J.O. Valderrama, Ind. Eng. Chem. Res. **42**, 1603 (2003).

3.2 J.S. Lopez-Echeverry, S. Reif-Acherman and E. Araujo-Lopez, Fluid Phase Equilib. **447**, 39 (2017).

3.3 M.S. Al-Manthari, M.A. Al-Wadhahi, G.R. Vakili-Nezhaad and K. Nasrifar, Int. J. Thermophys. **22**, 105 (2019).

3.4 U.K. Deiters and T. Kraska, *High-pressure fluid phase equilibria*, Elsevier, Oxford (2012).

3.5 S. Beret and J.M. Prausnitz, AIChE J. **21**, 1123 (1975).

3.6 O. Redlich and J.N.S. Kwong, Chem. Rev. **44**, 233 (1949).

3.7 G. Soave, Chem. Eng. Sci. **27**, 1197 (1972).

3.8 D.Y. Peng and D.B. Robinson, Ind. Eng. Chem. Fundam. **15**, 59 (1976).

3.9 N.C. Patel and A.S. Teja, Chem. Eng. Sci. **37**, 463 (1982).

3.10 A. Péneloux, E. Rauzy and R. Fréze, Fluid Phase Equilib. **8**, 7 (1982).

3.11 F.H. Ree and W.G. Hoover, J. Chem. Phys. **40**, 939 (1964).

3.12 N.F. Carnahan and K.E. Starling, J. Chem. Phys. **51**, 635 (1969).

3.13 G.S. Devendorf and D. Ben-Amotz, J. Phys. Chem. **97**, 2307 (1993).

3.14 U. Setzmann and W. Wagner, Int. J. Thermophys. **10**, 1103 (1989).

3.15 A. Saul and W. Wagner, J. Phys. Chem. Ref. Data **18**, 1537 (1989).

3.16 U. Setzmann and W. Wagner, J. Phys. Chem. Ref. Data **20**, 1061 (1991).

3.17 R.A. McDowell and F.H. Kruse, J. Chem. Eng. Data **7**, 547 (1963).

3.18 O. Kunz and W. Wagner, J. Chem. Eng. Data **57**, 3032 (2012).

3.19 R. Span and W. Wagner, Int. J. Thermophys. **24**, 41 (2003).

3.20 R. Span and W. Wagner, Int. J. Thermophys. **24**, 111 (2003).

3.21 D. Bücker and W. Wagner, J. Phys. Chem. Ref. Data **35**, 929 (2006).

3.22 M. Thol, R. Lustig, R. Span and J. Vrabec, J. Phys. Chem. Ref. Data **45**, 023101 (2016).

3.23 D.G. Friend, H. Ingham and J.F. Ely, J. Phys. Chem. Ref. Data **20**, 275 (1991).

3.24 J.E. Proctor, M. Bailey, I. Morrison, M.A. Hakeem and I.F. Crowe, J. Phys. Chem. B **122**, 10172 (2018).

3.25 http://thermoc.uni-koeln.de/ U.K. Deiters, Chem. Eng. Technol. **23**, 581 (2000).

3.26 M. Podsiadlo, A. Olejniczak and A. Katrusiak, Cryst. Growth Des. **17**, 228 (2017).

3.27 E.H. Brown, Bull. Intnl. Inst. Refrig., Paris, Annexe **1960–1961**, 169 (1960).

3.28 A. Neumaier and U.K. Deiters, Int. J. Thermophys. **37**, 96 (2016).

3.29 R. Span and W. Wagner, J. Phys. Chem. Ref. Data **25**, 1509 (1996).

3.30 https://webbook.nist.gov/chemistry/fluid/; https://www.nist.gov/srd/refprop

3.31 http://www.thermo.rub.de/en/prof-w-wagner/software/fluidcal.html

3.32 https://www.aspentech.com/en/products/engineering/aspen-hysys

3.33 http://www.ddbst.com/ddb.html

4

The Liquid State Close to the Melting Curve (I): Static Properties

In this chapter and Chapter 5 we will consider the drastic changes in the properties of the liquid state as we move away from the vapour pressure curve and approach the melting curve. We begin by noting that the density of liquids in this P, T area is nearly as high as that of the corresponding solid when the melting curve is crossed. In fact, in one notable exceptional case (H_2O) it is greater than the density of the solid. Therefore, the properties of liquids and dense supercritical fluids close to the melting curve have a lot in common with the properties of solids, as do the experimental methods used to study them. We make the division between consideration of static properties (this chapter) and dynamic properties (Chapter 5). However, such a division comes with a health warning. It is not possible to make consideration of the static and dynamic properties completely separate pursuits. Take the archetypal static property, density. If the temperature of a solid or dense fluid is greater than 0 K then the density is reduced, often significantly so, by the presence of phonons; the archetypal dynamic property. In the previous chapter we saw how both static properties (for example the P, V, T EOS) and dynamic properties (for example heat capacities) are derived from the same fundamental EOS.

4.1 Density and Bulk Modulus of Fluids Close to the Melting Curve

As a general rule, the density of a sample only decreases by 5 - 10% upon melting and the bulk modulus of the fluid close to the melting curve is also close to that of a solid. We illustrate this here with two examples (Ar and N_2, selected due to good data being available in the literature). For fluid N_2 we additionally calculate the bulk modulus of the fluid in the vicinity of the melting curve. Data are shown in Appendices A and B for other systems.

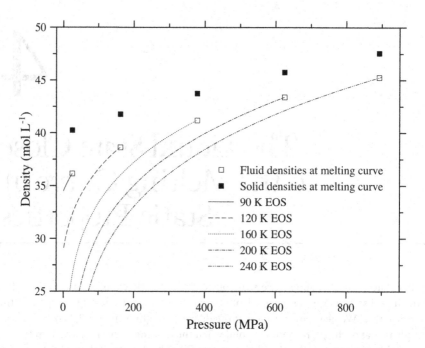

FIGURE 4.1 Fluid EOS (from NIST REFPROP) and volume changes upon melting (calculated from ref. [4.2]) for Ar from 90 K (close to the triple point of 83.8 K) to 240 K ($T/T_C = 1.59$).

4.1.1 Density of Fluid Ar Close to the Melting Curve

The fundamental EOS for fluid Ar available from NIST REFPROP is valid up to the melting curve at temperatures up to ca. 240 K ($T/T_C = 1.59$) [4.1]. It is in reasonable agreement with experimental measurements of the fluid density at the melting curve [4.2]. In Figure 4.1 the fluid EOS from NIST is plotted at various temperatures from close to the triple point (83 K) up to 240 K, along with the change in density upon solidification obtained from comparing the fluid and solid densities obtained in ref. [4.2]. The data is tabulated in Table 4.1. The density increase upon solidification is relatively small, decreasing from ca. 10% to ca. 5% upon temperature increase in the range studied.

TABLE 4.1 Fluid (ρ_F) and solid (ρ_S) densities of Ar at selected temperatures on the melting curve, and density change upon melting.

Temperature T (K)	T/T_C	ρ_F Mol L^{-1}	ρ_S Mol L^{-1}	Density Change $\dfrac{\rho_S - \rho_F}{\rho_S}$
90	0.60	36.15	40.25	0.102
120	0.80	38.63	41.77	0.075
160	1.06	41.20	43.74	0.0582
200	1.33	43.39	45.74	0.051
240	1.59	45.25	47.57	0.049

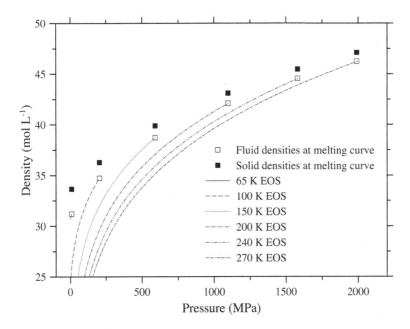

FIGURE 4.2   Fluid EOS (from NIST REFPROP) and volume changes upon melting (calculated from ref. [4.5]) for N_2 from 65 K (close to the triple point of 63.14 K) to 270 K ($T / T_C = 2.14$).

4.1.2 Density and Bulk Modulus of Fluid N_2 Close to the Melting Curve

The fundamental EOS [4.3] for fluid N_2 available from NIST REFPROP is valid up to the melting curve (the author has checked this against the experimental measurements made of the fluid density at the melting curve [4.4]). In addition, the density change of fluid N_2 upon melting has been measured (ref. [4.5] and refs. therein) from the triple point (63.151 K) up to 273 K, covering the entire temperature range in which the liquid exists and a considerable portion of the supercritical fluid region ($T_C = 126$ K). In Figure 4.2 the EOS at various temperatures in the liquid and supercritical fluid states is plotted up to the melting curve. As shown in the figure and tabulated data in Table 4.2, the density change when the melting curve is crossed is tiny. It declines from 7% at 65 K to 2% at 270 K.

In the temperature range considered here N_2 solidifies into the hexagonal close packed phase I [4.6]. This phase has *some* rotational order; the N_2 molecule is free to precess around a certain axis but consideration of the density alone precludes any model in which the molecule occupies a spherical space giving it freedom to rotate around any axis [4.7]. Considering the tiny density decrease upon melting, we may conclude that the liquid and supercritical fluid states of N_2 close to the melting curve could exhibit some short-range order. On the other hand, recent ab initio MD simulations of fluid N_2 at 300 K up to 2000 MPa [4.8] did not provide evidence for any short-range order.

TABLE 4.2   Fluid (ρ_F) and solid (ρ_S) densities of N_2 at selected temperatures, and density change upon melting, obtained from citations given in text.

Temperature T (K)	T/T_c	ρ_F Mol L^{-1}	ρ_S Mol L^{-1}	Density Change $\dfrac{\rho_S - \rho_F}{\rho_S}$
65	0.52	31.20	33.67	0.073
100	0.79	34.74	36.29	0.043
150	1.19	38.72	39.89	0.029
200	1.58	42.11	43.10	0.023
240	1.90	44.54	45.45	0.020
270	2.14	46.24	47.10	0.018

In both N_2 and Ar (and the other examples shown in Appendices A and B) the density change upon melting steadily decreases upon rising temperature. This has caused a lot of speculation over the decades about what happens to the melting curve upon further P,T increase. Does it ever end at a critical point in an analogous way to the vapour pressure curve? Does it exhibit a maximum? These issues are explored in Section 4.5.

In addition to consideration of the small density change upon melting it is instructive to study the density EOS of the fluid close to the melting curve. Visual inspection of Figures 4.1 and 4.2 suggests that at P,T conditions close to the melting curve the bulk modulus is high (low compressibility), especially when the melting curve is approached at higher P,T. But how high is it? How does it compare to solids? We may make a simple comparison by fitting a suitable solid-like EOS to the NIST data in this P,T region. The Murnaghan equation [4.9][4.10] (equation 4.1) is a simple invertible EOS widely used for solids. The volume V and pressure P are linked via two adjustable constants, the bulk modulus K_0 and its dimensionless pressure derivative K_0'. The volume at zero pressure is fixed at V_0.

$$\frac{V(P)}{V_0} = \left[1 + \frac{K_0' P}{K_0} \right]^{-\frac{1}{K_0'}} \tag{4.1}$$

To apply this to fluids close to the melting curve it is best to choose some reference conditions other than zero pressure since for a fluid $V \to \infty$ as $P \to 0$. So we will choose the fluid pressure and volume at the melting curve as P_0, V_0. The Murnaghan equation then becomes:

$$\frac{V(P)}{V_0} = \left[1 + \frac{K_0'(P - P_0)}{K_0} \right]^{-\frac{1}{K_0'}} \tag{4.2}$$

As an illustrative example, let us fit equation 4.2 to the N_2 NIST REFPROP data from 1500 MPa up to the melting curve at 270 K (Figure 4.3). The fit parameters obtained are:

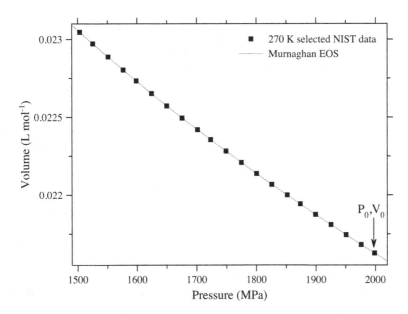

FIGURE 4.3 Plot of selected data points (squares) from NIST REFPROP fundamental EOS for N_2 at 270 K between 1500 MPa and the freezing point. Fit to these data using a Murnaghan EOS (dotted line, equation 4.2). P_0, V_0 are the parameters of the fluid when the melting curve is reached.

$K_0 = 8.9$ GPa and $K_0' = 4.4$. The value of K_0 is typical of what is observed at ambient conditions in more compressible elemental solids (e.g., Ba, Li, Na, S). In addition, the value of K_0' lies at a value typical for a solid.

4.2 Elastic Neutron and X-ray Diffraction from Liquids Close to the Melting Curve

A large amount of information on the structure of fluids has been obtained from scattering of neutrons and X-rays. Such experiments are performed even on gases; in Figure 1.1 we saw data from gaseous Ar in which a structure factor $S(Q)$ with a clearly measurable first peak is present at 19.3 MPa (190 bars). The author once collected diffraction data at the ISIS pulsed neutron source on N_2 at 25 MPa which also clearly showed a first peak in $S(Q)$ [4.8]. However, at densities as low as this the data collection time can be exorbitantly long and the poor signal-to-noise ratio prohibits detailed analysis by Fourier transform or empirical potential structure refinement. Hence neutron and X-ray diffraction are used principally to study fluids at higher densities close to the melting curve. We discuss elastic scattering in this section (that we will refer to as diffraction) and inelastic scattering in Chapter 5, whilst noting that describing the scattering as completely elastic is an approximation ([4.11] and references therein).

The methods utilized to analyse fluid diffraction data are the same as those for diffraction data from amorphous solids, and some analogies can even be drawn between analysis of diffraction data on fluids and on crystalline solids. In both cases, the diffraction data can be described mathematically as the Fourier transform of the real-space structure of the sample. That is, the structure factor $S(\mathbf{Q})$ is the Fourier transform of the function $\rho(\mathbf{r})$ representing the structure of the sample in real space (see proof in Appendix C).

However, there are limitations to this analogy. It does not generally extend to EOS measurements. X-ray diffraction is the most widely used and accurate method to obtain P,V,T EOS data on crystalline solids over a wide P,T range. The author has spent a large portion of his career compressing solids to pressures up to 200 GPa in the diamond anvil cell and obtaining the EOS from diffraction data.

Fluid diffraction data is not generally used to obtain the P,V,T EOS, though this has been attempted [4.12]. The difficulty in applying this approach to fluids is not just because the fluid diffraction peaks are weaker and broader. When the structure of a solid is known it is acceptable to obtain the density from the diffraction data by assuming that all atoms are present and correct in the unit cell. This assumption cannot be made for a fluid since the absolute co-ordination number and the density of vacancies in the fluid cannot generally be calculated with sufficient accuracy from diffraction data. None of the fluid EOS data discussed in this book are obtained by diffraction and this method should only be used when no other data are available. We outline in Section 4.2.2 how fluid EOS measurement using diffraction is performed.

To obtain more general information about the structure of a fluid (or a solid for that matter) from diffraction data there are essentially two approaches. In the first approach, the $S(Q)$ data are Fourier transformed to obtain $g(r)$ [4.12][4.13]. If we wish, we can compare this to a $g(r)$ obtained from a simulation of the structure of the fluid. In the second approach [4.14][4.15], a simulated structure is Fourier transformed to obtain a simulated $S(Q)$ and this is compared to the experimentally obtained $S(Q)$. The simulated structure is adjusted to produce an $S(Q)$ most closely resembling the experimental data.

In an ideal world both approaches would be mathematically equivalent, and both would work perfectly. This is not the case in practice. The raw data obtained in a diffraction experiment is not $S(Q)$. To obtain $S(Q)$ from the raw diffraction data it is necessary to subtract a background signal. The background signal may well be as strong a signal as the data and not straightforward to accurately measure, for instance because part of the background will be the incoherent scattering from the sample. Then it is necessary to make a number of other corrections (e.g., for detector efficiency). In addition, in the case of X-ray diffraction the drop in the magnitude of the atomic form factor for higher Q results in the obtained $S(Q)$ being incomplete (covering only lower values of Q). If (as is always the case, even for neutron diffraction) the data are not available to $Q \to \infty$, the Fourier transform is no longer exact.

The fundamental question is—which approach to analysis of diffraction data is affected less by these errors? We will review examples where both approaches have been successful. However, it is necessary to begin by discussing the differences between X-ray and neutron diffraction experiments.

4.2.1 Distinctions Between X-ray and Neutron Diffraction Experiments

It is useful to understand various experimental and theoretical distinctions between X-ray and neutron diffraction. In both cases, the diffraction pattern is a result of the interference between beams scattered coherently by each individual atom in the sample. However, the nature of the scattering of X-rays and neutrons by the individual atoms is fundamentally different. Neutrons are scattered solely by the nucleus, with which they interact solely via the strong nuclear force. X-rays are scattered by the electrons in the atoms, with which they interact solely via the electromagnetic force.

To obtain a diffraction pattern the wavelength of the neutrons and X-rays has to be of the same order of magnitude as the spacing between adjacent atoms and therefore the same order of magnitude as the size of the atom. Therefore, scattering of coherent X-rays is attenuated at large Q (which corresponds to large scattering angle θ) due to the fact that there is a large optical path difference between beams scattered from opposite sides of the same atom. This phenomenon is encapsulated in the parameter of the atomic form factor $f(Q)$. The coherent scattering intensity from the single atom is given by $f(Q)^2$. Figure 4.4 illustrates this in the context of the diffraction experiment geometry and shows $f(Q)^2$ for carbon plotted from tabulated data in the literature [4.16]. The parameter Q (the modulus of the scattering vector) is simply the change in wavevector upon diffraction. Using the notation from Figure 4.4 (b) we can write:

$$Q = k' - k$$

The scattering angle is defined as 2θ for consistency with the Bragg formula, leading to:

$$Q = |k' - k| = \frac{4\pi \sin \theta}{\lambda}$$

For molecular fluids, destructive interference between beams scattered from opposite sides of the same molecule can cause further attenuation at large Q in an analogous manner. CH_4 is an extreme example—due to the spherical symmetry of the molecule it essentially acts as a large noble gas atom and one can define a "molecular form factor" to account for this. The author once conducted synchrotron X-ray diffraction experiments on fluid CH_4 in which only the first peak in the structure factor $S(Q)$ was observed with significant intensity due to this attenuation [4.17]. We should have known better!

In contrast, the intensity of neutron scattering is independent of Q so a neutron diffraction experiment can obtain data at much higher values of Q. The only constraint on Q is from the requirement that $\theta < 180°$. On the other hand, the interaction of neutrons with matter solely via the strong nuclear force results in neutron diffraction being, overall, a very weak effect. An incident neutron passes through a very large number of atoms before coming close enough to a nucleus to interact with the nucleus via the short-ranged strong nuclear force. As a result, data collection times when performing neutron diffraction on fluids are many hours, whilst it takes a few minutes at a modern synchrotron X-ray source. Mapping out fluid properties over a large P,T range is far more feasible using X-rays.

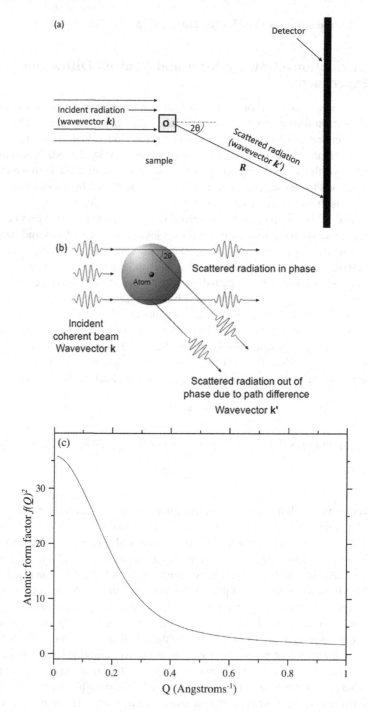

FIGURE 4.4 (a) Geometry for a diffraction experiment. (b) Elastic scattering of a coherent X-ray beam from a single atom (credit: Dr. Daniel Alfonso Melendrez Armada). X-rays scattered through a large angle from different parts of the atom are out of phase so interfere destructively. Reproduced with permission from ref. [4.18]. (c) Plot of $f(Q)^2$ for carbon using tabulated data from ref. [4.16].

The diffraction of X-rays from the electron clouds and diffraction of neutrons from the nucleus also ensures that the variation of scattering intensity between different elements and isotopes is fundamentally different. The X-ray scattering intensity varies smoothly in proportion to Z^2 and is the same for all isotopes of an element, whilst the neutron scattering intensity varies haphazardly as a function of Z and varies drastically for different isotopes of the same element. This is because it depends on the nature of the nucleus and is not directly related to the charge on the nucleus. This means that in polyatomic fluids it is often possible to obtain significantly more information by conducting the experiment with different isotopes of the same element, a technique called isotopic substitution. In the extreme and important case of hydrogen-rich materials, conducting the diffraction from a sample with hydrogen and then from a separate sample with deuterium can be extremely useful to obtain separate sets of information about the distribution of hydrogen (deuterium) in the sample and about the distribution of the remaining elements in the sample. This is due to the coherent scattering intensity for deuterium being far larger than that for hydrogen.

Finally, in both neutron and X-ray diffraction it is often appropriate to focus the beam to a small spot size. A smaller sample allows higher pressures to be safely reached, and this necessitates a smaller spot size as the sample is often unavoidably surrounded by materials which scatter neutron and X-ray beams strongly. However, the beam cannot be smaller than the diffraction limit imposed by the uncertainty principle. This would be 1–10 nm beam diameter for both neutrons and X-rays at the wavelengths utilized for diffraction experiments. We have the technology to focus X-rays to a spot size comparable to this diameter, and 1 µm spot size is achieved routinely at synchrotron X-ray sources around the world [4.19]. The technology does not exist to do this with neutrons—neutron spot sizes are typically a few mm, leaving a challenge for future generations of engineers.

4.2.2 Fourier Transform of Fluid Diffraction Data to Obtain $g(r)$

Techniques to Fourier transform diffraction data to obtain $g(r)$ and other parameters have been outlined in various publications over the years, for instance in ref. [4.12]. The mathematical method begins with the Debye scattering equation (equation 4.3). We will follow the methodology of refs. [4.12][4.20] but with slightly different definitions/notations to allow consistency with discussion on diffraction elsewhere in this text:

$$I_{coh}(Q) = \sum_{m=1}^{N} \left[\sum_{n=1}^{N} f_m(Q) f_n(Q) \frac{\sin(Qr_{mn})}{Qr_{mn}} \right] \tag{4.3}$$

The Debye scattering equation sums the coherent scattering amplitude and phase factor for N atoms in a disordered sample to obtain the total amplitude of coherent scattering $I_{coh}(Q)$. Q is the scattering vector, directly related to the scattering angle. The lack of long-range order in the fluid is assumed to result in the scattering being spherically symmetric. $f(Q)$ indicates the atomic form factors of the atoms and r_{mn} the distance between atoms m and n. We will consider the case of neutron diffraction from a monatomic fluid for simplicity. In this case all atoms have the same constant atomic form factor $f(Q) = b$, which we refer to as the scattering length. However, the conceptual results can be applied to X-ray scattering and to polyatomic/molecular fluids also. Using symmetry considerations for very large N the second summation takes the same value for every value of m so we can simplify to:

$$I_{coh}(Q) = Nb^2 \sum_{n=1}^{N} \frac{\sin(Qr_n)}{Qr_n} + Nb^2$$

The first term accounts for the addition of scattered amplitude from a single, centrally located atom and each other atom in the fluid, whilst the second term corresponds to all the terms in equation 4.3 for which $n = m$. In a real X-ray or neutron diffraction experiment the beam diameter is sufficiently large that it is a reasonable approximation to state that the sample (illuminated by the beam) is infinitely large compared to the interatomic spacing. We therefore convert the summation into an integral using the radial distribution function $g(r)$ and a volume element around the surface of a sphere of radius r. Therefore, $g(r) \times 4\pi r^2 dr$ is the number of atomic centres in this volume element. We obtain:

$$I_{coh}(Q) = Nb^2 \left[1 + \int_{r=0}^{r=\infty} g(r) \frac{\sin(Qr)}{Qr} 4\pi r^2 dr \right]$$

We then separate the scattering into terms that are dependent and independent of r by adding and subtracting a constant atomic density g_0 to obtain:

$$I_{coh}(Q) = Nb^2 \left[1 + \int_{r=0}^{r=\infty} [g(r) - g_0] \frac{\sin(Qr)}{Qr} 4\pi r^2 dr + \int_{r=0}^{r=\infty} g_0 \frac{\sin(Qr)}{Qr} 4\pi r^2 dr \right]$$

The first term in this expression is a constant which can be subtracted from the experimentally observed intensity and the last is scattering at extremely small angles (small Q) such that it is blocked along with the unscattered beam. We then define the structure factor $S(Q)$ as the scattered intensity normalized in the high-Q limit (equation 4.4) which leaves a single integral (equation 4.5):

$$S(Q) = \frac{I_{coh}(Q)}{Nb^2} \tag{4.4}$$

$$S(Q) - 1 = \int_{r=0}^{r=\infty} [g(r) - g_0] \frac{\sin(Qr)}{Qr} 4\pi r^2 dr \tag{4.5}$$

Performing a Fourier sine transform[*] on this integral leads to:

$$4\pi r [g(r) - g_0] = \frac{2}{\pi} \int_{0}^{\infty} Q[S(Q) - 1] \sin(Qr) dQ \tag{4.6}$$

---

[*] The Fourier sine transform can be performed only on an odd function. Therefore, $[g(r) - g_0]$ has to be even. This is acceptable, since we can choose whatever behaviour we wish the function to have for $r < 0$ without affecting the value of the integral.

This Fourier transform relationship is exact as written and therefore, as long as the background to the diffraction signal is known precisely, $g(r)$ can be calculated accurately from $I_{coh}(Q)$ via $S(Q)$. The calculation of $g(r)$ is possible with reasonable accuracy in a neutron diffraction experiment since data are typically available out to extremely high Q due to the atomic form factor being a constant, b.

However, in X-ray diffraction the constant parameter b is replaced with the Q-dependent parameter of the atomic form factor $f(Q)$ and $S(Q)$ is calculated from the coherent scattering intensity $I_{coh}(Q)$ via $I_{coh}(Q) = Nf(Q)^2 S(Q)$. The measured scattering intensity therefore drops rapidly to zero with increasing Q due to the trend in $f(Q)^2$ illustrated in Figure 4.4 (c), regardless of the value of $S(Q)$. Since $S(Q)$ at high Q cannot be calculated we cannot calculate $g(r)$ using its exact relationship to $S(Q)$ given in equation 4.6. The simplest remedy to this problem is to truncate the integral at a maximum value, Q_{max}, at which $f(Q_{max})^2 \approx 0$. This gives:

$$4\pi r\left[g(r) - g_0\right] = \frac{2}{\pi} \int_0^{Q_{max}} Q\left[S(Q) - 1\right]\sin(Qr)\,dQ$$

The result of this approximation is drastic changes to the higher r part of $g(r)$ that depend sharply on where that cutoff is chosen. Figure 4.5 demonstrates an example

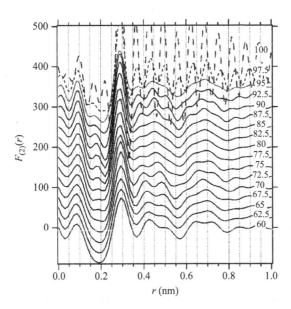

FIGURE 4.5 Variations in the obtained X-ray $g(r)$ $\left(F_{(2)}(r) = 4\pi r\left[g(r) - g_0\right]\right)$ of liquid H_2O as a function of the cutoff value chosen for Q (values listed on the right in nm^{-1}) [4.12]. The variations are particularly pronounced at high r. Reprinted figure with permission from J.H. Eggert, G. Weck, P. Loubeyre and M. Mezouar, Phys. Rev. B **65**, 174105 (2002), https://doi.org/10.1103/PhysRevB.65.174105. Copyright (2002) by the American Physical Society.

of this from ref. [4.12]. Significant changes to all parts of the $g(r)$ function occur at all r as the cutoff is varied, although the changes at low r are less drastic. This problem of the cutoff of X-ray diffraction data at high-Q is widely discussed in the literature [4.21][4.22][4.23][4.24]. In Sections 4.2.3 and 4.2.4 we discuss two approaches taken to resolve this problem that are more sophisticated than simply truncating the integral.

There is, however, useful information that can be obtained from the Fourier-transformed data, even in the case of X-ray diffraction. For molecular fluids, the low-r part of $g(r)$ contains information on the intra-molecular bond(s) and the high-r part contains information on the inter-molecular bonds/distances. As we have seen, the low-r part can be more reliably obtained (in Figure 4.5 the changes to $g(r)$ below $r \approx 0.15$ nm are minimal). It can also be modelled accurately using the known atomic form factors, and in the case of molecular fluids using the intra-molecular bond lengths known from the gas phase. It is a reasonable assumption that these do not change significantly until the melting curve is approached. The adjustable fitting parameters used in this modelling are the average density ρ_0 and scaling factors related to the subtraction of the background and the incoherent scattering signal from the sample. It is thus, despite the fact that most of the $g(r)$ function cannot be obtained reliably, possible to measure the density of a fluid using X-ray diffraction.

Density and hence P,V EOS measurement using X-ray diffraction is not a widely used technique; potentially the large errors that can result from the slightest inaccuracy in background subtraction are a factor in this. The P,V equations of state of Ar and H_2O have been measured with this method to a reasonable accuracy (agreeing with conventional direct measurements of volume within ±3%) [4.12] (Figure 4.6). However, it is not certain that this accuracy is achieved in all EOS calculations using this method. On the other hand, the advantage of the X-ray diffraction method is that it can be utilized at very high pressures. It has been employed at megabar pressures [4.25], roughly two orders of magnitude higher than the pressures at which direct measurements of sample volume are possible. Interestingly, P,V EOS measurement using the combination of Brillouin scattering and refractive index measurements has also been attempted under extreme conditions in the DAC (diamond anvil cell) [4.26].

4.2.3 Fourier Transform of Modified Fluid Diffraction Data to Obtain $g(r)$

Over the decades, beginning with the paper of Lorch in 1969 [4.27], attempts have been made with some success to find mathematical methods ameliorating the Q_{max}-cutoff problem that was described in the previous section. Opinions differ on how valid these methods are [4.21]; the fact that a method suppresses unphysical fluctuations in $g(r)$ at large r or produces the expected results does not prove that it is necessarily correct. Below we outline how the controversy arises, from the mathematical point of view. Following this it is possible to understand how Lorch's method works and the conditions under which it works.

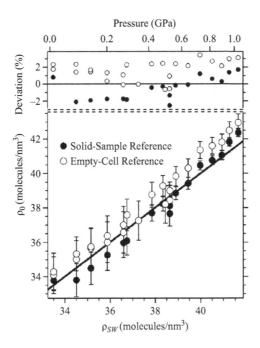

FIGURE 4.6 Plot of the density of liquid H_2O measured using X-ray diffraction (*y*-axis) against the density according to the fundamental EOS (*x*-axis) [4.12]. Filled circles correspond do densities obtained following subtraction of a background signal from a solid sample in the DAC (i.e., ice), and open circles to subtraction of a background signal from an empty DAC. Reprinted figure with permission from J.H. Eggert, G. Weck, P. Loubeyre and M. Mezouar, Phys. Rev. B **65**, 174105 (2002), https://doi.org/10.1103/PhysRevB.65.174105. Copyright (2002) by the American Physical Society.

Lorch began with the exact Fourier transform relation between $S(Q)$ and $g(r)$ exactly as derived earlier (equation 4.6), reproduced below. Where Lorch used the notation $i(Q)$ we use $S(Q)$, which is obtained from the coherent diffracted intensity $I_{coh}(Q)$ normalized in the high-Q limit (equation 4.4).

$$4\pi r^2 \left[g(r) - g_0 \right] = \frac{2}{\pi} \int_0^\infty Qr \left[S(Q) - 1 \right] \sin(Qr) dQ$$

Lorch's method is essentially to make a compromise. Instead of seeking to determine the exact value of $g(r)$ for all r, we obtain just the integrated value of $g(r)$ over a range of values about some central value r_0: $r_0 - \Delta \leq r \leq r_0 + \Delta$ (Lorch used $\Delta/2$ instead). In this case, both sides of equation 4.6 are integrated with respect to r within these limits. Equation 4.7 shows this integral.

$$\int_{r_0-\Delta}^{r_0+\Delta} 4\pi r_0^2 \left[g(r) - g_0 \right] dr = \frac{2}{\pi} \int_{r=r_0-\Delta,Q=0}^{r=r_0+\Delta,Q=\infty} Qr \left[S(Q) - 1 \right] \sin(Qr) dQ \, dr \qquad (4.7)$$

In Lorch's original paper, the assumption $\Delta \ll r_0$ is stated but it is not explained where the assumption must be applied and what the consequences are if we are not on solid ground with this assumption. Here these matters are outlined. Dealing with the left-hand side (LHS) first, we would like to continue referring to the non-integrated value of $g(r)$. To achieve this, we assume that $\Delta \ll r_0$ and that $g(r)$ does not vary significantly over the region covered by integral over r. Equation 4.8 below specifies the approximation we make:

$$\int_{r_0-\Delta}^{r_0+\Delta} 4\pi r_0^2 \left[g(r) - g_0 \right] dr \approx 2\Delta \times 4\pi r_0^2 \left[g(r_0) - g_0 \right] \tag{4.8}$$

We then apply the $\Delta \ll r_0$ assumption to the right-hand side (RHS) of equation 4.7. We assume that Δ is sufficiently small that the variation in the value of the first instance of r is negligible over the range of the integral. We therefore set $r \rightarrow r_0$ here. It does *not* follow from this that we should set $\sin(Qr) \rightarrow \sin(Qr_0)$ also, because the integral extends to $Q = \infty$ in which case this term oscillates very rapidly as a function of r. We therefore need to perform the integral with respect to r in the equation below:

$$2\Delta \times 4\pi r_0^2 \left[g(r_0) - g_0 \right] = \frac{2}{\pi} \int_{r=r_0-\Delta, Q=0}^{r=r_0+\Delta, Q=\infty} Qr_0 \left[S(Q) - 1 \right] \sin(Qr) dQ dr$$

We integrate, applying standard trigonometric identities[†] and obtain:

$$4\pi r_0^2 \left[g(r_0) - g_0 \right] = \frac{2}{\pi} \int_0^{Q_{max}} Qr_0 \left[S(Q) - 1 \right] \sin(Qr_0) \left[\frac{\sin(Q\Delta)}{Q\Delta} \right] dQ \tag{4.9}$$

The result of Lorch's compromise is an expression on the RHS of equation 4.9 identical to equation 4.7 except that the integrand is now multiplied by the sinc function (in the square brackets). Since this decays rapidly to zero upon increasing Q, it is no longer necessary to evaluate the integral to $Q = \infty$ and we can instead insert the upper limit $Q = Q_{max}$. The lower the resolution we require in our evaluation of $g(r)$, the smaller Q_{max} can be; from the properties of the sinc function we obtain $Q_{max} \approx \pi / \Delta$ since the function has decayed to a very small value by this point. Equation 4.9 has been used as a standard method to address the Q_{max}-cutoff problem for the 50 years since Lorch's paper was published; it has 613 citations to date, 164 of which are from 2015 onwards. The method has been used in many papers fundamental to our understanding of fluids, glasses, and amorphous solids [4.13][4.28][4.29][4.30][4.31][4.32] [4.33][4.34].

[†] $\cos(A+B) = \cos A \cos B - \sin A \sin B$ etc.

However, if there is any problem with the $\Delta \ll r_0$ assumption Lorch's method fails catastrophically. If we integrate the RHS of equation 4.7 without leaving the first instance of r constant, we obtain equation 4.10, in agreement with ref. [4.21].

$$4\pi r_0^2 \left[g(r_0) - g_0 \right] = \frac{2}{\pi} \int_0^{Q_{max}} Q r_0 \left[S(Q) - 1 \right] \sin(Q r_0) \left[\frac{\sin(Q\Delta)}{Q\Delta} \right] dQ$$

$$+ \frac{2}{\pi} \int_0^{\infty} \left[S(Q) - 1 \right] \left[\frac{\sin(Q\Delta)\cos(Q r_0)}{Q\Delta} - \cos(Q r_0)\cos(Q\Delta) \right] dQ$$

(4.10)

In this case there can be no justification for setting an upper limit below $Q = \infty$ for the second integral, which clearly does not have a small value compared to the first integral.

Therefore we must conclude that Lorch's method can only be used when the radius r_0 about which we wish to calculate $g(r)$ is large compared to Δ. The mathematical failure of the Lorch method when this condition is not met is potentially catastrophic. Since the smallest possible value of Δ is determined by Q_{max}, which is in turn determined by the maximum scattering angle and/or atomic form factor, there is a clear constraint on the minimum value of r_0 for which we can obtain $g(r)$ accurately using the Lorch method. Combining conditions given earlier, this minimum value is given by:

$$r_0 \gg \frac{\pi}{Q_{max}}$$

(4.11)

For a typical neutron diffraction experiment $Q_{max} \approx 20$ Å^{-1} so the minimum value of r_0 is small compared to the radius of an atom, but for an X-ray diffraction experiment $Q_{max} \approx 5$ Å^{-1} is common so the minimum value of r_0 could easily lie within the first co-ordination shell and extreme care is required when applying the Lorch modification function to the data. In the original paper from 1969 [4.27], Lorch applied his function only to neutron diffraction data, though since then many have applied it to X-ray diffraction data.

4.2.4 Comparison of Diffraction Data to Simulated Fluid Structures in Reciprocal Space

Alternatively, one may compare the diffraction data to a simulated structure in reciprocal space. In this case, a background of comparable or greater intensity than the data must still be subtracted; no analysis method can eliminate all the potential errors arising from this subtraction. However, if we perform a Fourier transform of the diffraction data as outlined in the previous sections (with or without the Lorch modification) we have to deal with the additional uncertainty of not knowing how these potential errors, and noise in the data, will propagate through the Fourier transform (equation 4.6 or equation 4.9). This specific source of error can be avoided by comparing the diffracted data to simulated fluid structures in reciprocal space, as outlined in this section.

The simulation of the structure is usually done using a Monte Carlo method, most commonly empirical potential structure refinement (EPSR) [4.15]. The behaviour of a box of particles (typically several hundred) is simulated subject to certain constraints set by the user. The constraints set are typically a fixed density, certain atomic radii (to prevent overlap between adjacent atoms), the nature of the pair potential $\mathcal{V}(r)$ through which adjacent particles interact (e.g., Lennard-Jones), etc. The minimal constraints set allow the system enormous freedom to find a configuration that produces a structure factor most closely matching the experimental data, as described in the following paragraphs. EPSR and similar methods are therefore not comparable to simulation methods (for instance ab initio MD) designed for simulation of fluid properties without recourse to experimental data and are only suitable for conducting simulations to refine using experimental data. Monte Carlo methods developed prior to EPSR had even fewer constraints; no potential energy function was utilized but atomic overlap was still prevented. So in effect a potential that was $+\infty$ for $r < a$ and zero for $r \geq a$ was being utilized [4.35].

In the EPSR simulation, different atomic movements are considered; whole particle movements for all fluids and intra-molecular rotations etc. for molecular fluids. If the move reduces the overall potential energy of the system (for instance, if it brings unlike charges closer together) it is always accepted. If the move increases the overall potential energy of the system (e.g., by bringing like charges closer together) the probability of the move being accepted is given by the Boltzmann factor:

$$e^{-\frac{\Delta U}{k_B T}} \tag{4.12}$$

Here, ΔU is the potential energy change.

At each stage of the simulation the radial distribution function $g(r)$ can be calculated from the atomic positions and Fourier transformed to obtain $S(Q)$ for comparison to the diffraction data. The empirical potential (EP, U_{EP}) is the method by which this information is utilized. The overall methodology is as follows. Initially the simulation is run without comparison to the diffraction data. The potential that is minimized is the function giving the total potential of the system as a function of the atomic positions. This has a mathematical form such as the Lennard-Jones potential that is set *a priori* by the user. This potential is referred to as the "reference" potential U_{REF}. The simulation is then run with a different potential energy U minimized (equation 4.13); U_{REF} plus the EP.

$$U = U_{REF} + U_{EP} \tag{4.13}$$

At each stage the diffraction pattern(s) from the system are simulated and compared to the available diffraction data. The EP quantifies how well the simulated data agrees with the actual data with a larger EP indicating poorer agreement. There are two crucial remarks to make about this stage. Firstly, the simulated structure in real space is Fourier transformed into reciprocal space rather than the other way round. Therefore the Q_{max} cutoff problem described earlier is avoided. Secondly, many different diffraction datasets (the more, the better) can be utilized in the refinement process. The EPSR process can simulate neutron diffraction and X-ray diffraction data from the same sample and can simulate neutron diffraction data from samples of different isotopic composition.

Typically, EPSR takes many iterations to converge to a sensible result. The state which the system will converge to is a compromise (hopefully, but not necessarily, the best compromise) between the state which minimizes the interatomic potential energy with mathematical form set by the user, and the state which gives the best agreement between the simulated and experimental diffraction data.

Figure 4.7 (a) shows an example of an $S(Q)$ obtained experimentally by the author at the ISIS pulsed neutron source (fluid N_2 at 300 MPa and 300 K) alongside the $S(Q)$ simulated using EPSR following refinement. In this case a radial distribution function $g(r)$ (Figure 4.7 (b)) is obtained from the same simulation without the need to compare data to simulation in real space.

4.2.5 Relation Between $g(r)$, the Partition Function, Internal Energy, and Pressure

Above we have discussed how to extract the radial distribution function $g(r)$ as accurately as possible from diffraction data, so it is worthwhile also to understand its fundamental importance. The function is experimentally observable but can also be extracted directly from the partition function. Unfortunately, it is only possible to use this method to obtain the exact mathematical form of $g(r)$ in the low density limit since it is only in this limit that we can actually obtain the partition function. For a system of spherical particles [1.11] we obtain:

$$\lim_{\rho \to 0} g(r) = e^{-\frac{V(r)}{k_B T}} \tag{4.14}$$

Where $V(r)$ is the pair potential. Importantly, in this case $g(r)$ still has a substantial peak corresponding to some clustering of the particles. The physical cause of this is the formation of dimers due to the weak attractive van der Waals forces between particles, even for noble gas fluids. Mathematically it results from the inclusion of the pair potential $V(r)$ in equation 4.14 which is attractive close to some equilibrium separation r_0. This peak in $g(r)$ only disappears in the ideal gas limit, that is $k_B T \gg -V(r_0)$. Figure 4.8 shows the Lennard-Jones potential (equation 2.4) plotted alongside $g(r)$ obtained using equation 4.14 in the low density and ideal gas limits.

We may also relate $g(r)$ to the internal energy, pressure, and compressibility K of the system. Utilizing the pair potential $V(r)$, we can write [1.11][3.4]:

$$U = \frac{3}{2} N k_B T + 2\pi N \rho \int_{r=0}^{\infty} V(r) g(r) r^2 dr$$

$$\frac{P}{\rho k_B T} = 1 - \frac{2\pi\rho}{3k_B T} \int_{r=0}^{\infty} \frac{dV}{dr} g(r) r^3 dr$$

$$\rho k_B T K = 1 + 4\pi\rho \int_{r=0}^{\infty} \left[g(r) - 1 \right] r^2 dr \tag{4.15}$$

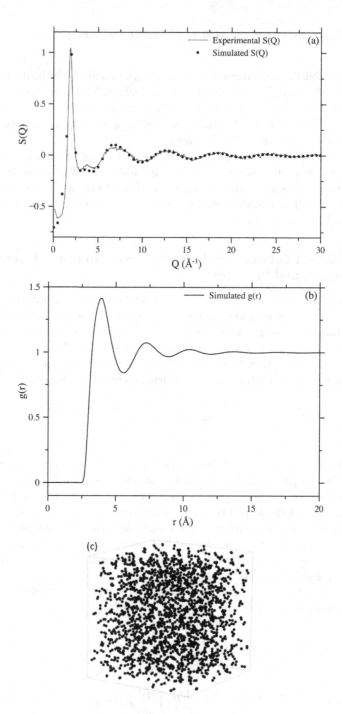

FIGURE 4.7 (a) Experimental structure factor $S(Q)$ of fluid N_2 at 300 MPa and 300 K (line) and selected points from an $S(Q)$ produced from an EPSR simulation refined to fit the data. (b) Radial distribution function $g(r)$ produced with the same simulation and (c) the box of 1000 molecules following simulation. Atomic diameters are set to 30% of the van der Waals radius for clarity; in reality the sample is a relatively closely packed structure at 300 MPa.

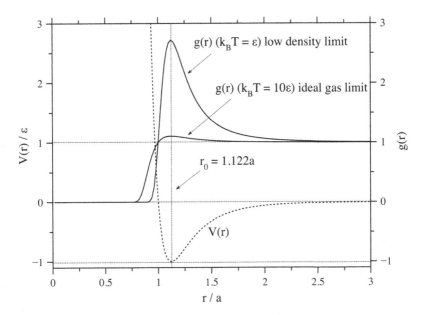

FIGURE 4.8 Lennard-Jones potential (dotted line, cf. Figure 2.1 and equation 2.4) plotted alongside $g(r)$ resulting from this potential in the low density and ideal gas limits (solid lines).

These widely used equations are referred to as the "energy equation," "pressure equation," and "compressibility equation." They are exact as written for spherical particles. Hence, if we know ρ, T and $V(r)$ then $g(r)$ can give us the internal energy U, pressure P, and compressibility K. To derive these equations the classical approximation has been used to divide the expressions into terms depending on the kinetic and potential energy.

4.2.6 Relation Between $g(r)$ and Entropy

The radial distribution function $g(r)$ can even be used to make a reasonable estimate of the configurational entropy of the fluid [4.36]. In this approach, the configurational entropy is the translational entropy (the same as that of an ideal gas under the specified ρ, T conditions) plus a term dependent on $g(r)$ representing two-particle interactions plus higher order terms dependent on the higher order distribution functions. The entropy contribution from the ideal gas and two-particle interactions is given by (in units of R):

$$S = \frac{5}{2} + \ln\left[\frac{1}{\rho}\left(\frac{2\pi m k_B T}{h^2}\right)^{\frac{3}{2}}\right] + 2\pi\rho\left[\int_{r=0}^{\infty}\left[g(r)-1\right]r^2 dr - \int_{r=0}^{\infty}g(r)\ln\left[g(r)\right]r^2 dr\right]$$

Higher order terms representing three-particle interactions would have to be added for this method to produce reasonable results at higher densities. The above expression is for a monatomic fluid; however, this method has been applied to the difficult molecular fluid example of H_2O [4.37].

4.2.7 Relation Between $g(r)$ and Co-ordination Number (CN)

The co-ordination number (CN) of the fluid can be calculated from $g(r)$. By definition, the number of particles within a spherical shell spanning distance r to $r+dr$ from a marked particle is $4\pi r^2 \rho g(r)dr$. Therefore the (nearest neighbour) co-ordination number is obtained by integrating $g(r)$ up to the first minimum (labelled by r_{min} in equation 4.16). Whilst the relation is mathematically exact, the calculation is subject to potential systematic error due to factors related to the Fourier transform procedure used to obtain $g(r)$ and in the exact location of r_{min}. Some studies have obtained unphysically large co-ordination numbers (above the close-packing limit of 12) [4.13][4.38] using this method and there is thus a consensus that this can only be relied upon to accurately calculate the change in co-ordination number due to changing P,T conditions in the same study [4.8][4.38]. Figure 4.9 shows an inter-molecular $g(r)$ function produced from refinement of neutron diffraction data on fluid N_2 at 300 K, 100 MPa using EPSR. The area obtained from the integral in equation 4.16 is shaded.

$$CN = 4\pi\rho \int_0^{r_{min}} r^2 g(r)dr \qquad (4.16)$$

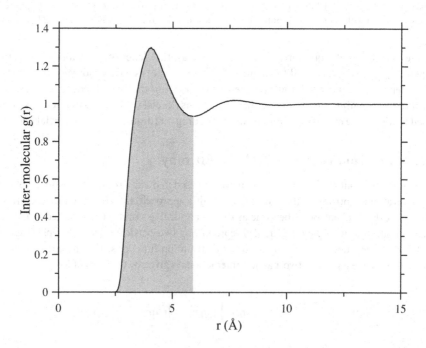

FIGURE 4.9 Inter-molecular $g(r)$ function for fluid N_2 at 300 K, 100 MPa produced from refinement of neutron diffraction data obtained by the author on the SANDALS diffractometer at the ISIS pulsed neutron source. The shaded area is the integral performed to obtain the co-ordination number (CN).

4.3 Short-Range Order and Phase Transitions in Fluids Close to the Melting Curve

In our discussion of fluid N_2 earlier we concluded that in the vicinity of the melting curve the fluid probably exhibits short-range (orientational) order, due to the fluid density being close to that of the orientationally ordered solid phase produced upon crystallization. There is simply not enough room for each N_2 molecule to occupy a spherical space. The same argument can be made for many other molecular fluids. C_2H_6 (ethane) is another good example—upon isothermal compression at 300 K, by ca. 1400 MPa there is not even enough space for each molecule to occupy a rod-shaped volume with the length of the molecule, let alone a spherical volume. Thus, from this point up to solidification at 2500 MPa there must be a high level of short-range order, as well as compression of the intra-molecular bonds to achieve the expected density [3.24]. Even in monatomic fluids, short-range structural order must exist close to the melting curve to achieve the observed densities and co-ordination numbers.

If short-range order exists, then the possibility arises that another solid-like phenomenon can be observed, phase transitions between configurations with different short-range order. In this section we will discuss representative examples of liquids exhibiting short-range order and, in some cases, phase transitions.

4.3.1 Co-ordination Number

Even a monatomic fluid can exhibit a high degree of short-range order close to the melting curve. Imagine compressing a sample of hard spheres—upon sufficient compression they will adopt the structure with the most efficient packing arrangement where each particle is surrounded by 12 neighbouring particles. Experimentally, it is observed that this level of order can be achieved, over short distances, at significantly lower pressure than that required for crystallization. Examples studied include N_2 [4.8], Ne [4.13], and H_2O [4.38].

4.3.2 Liquid-Liquid Phase Transitions

The study of liquid-liquid phase transitions is a field with a long history [4.39]. Here, we can begin by classifying the observed phase transitions into two categories. The first category is phase transitions in which the nature of the particles that comprise the fluid changes. Such a phase transition can alternately be understood as a chemical reaction or decomposition and can be observed in samples in the gas, liquid, or solid states. The second category is transitions in which the individual particles do not change, but some aspects of the arrangement of the particles changes. Transitions in this second category are observed far less frequently and can only be observed in samples with density high enough to exhibit short-range order in a manner analogous to a solid.

Good examples in the first category are sulphur and phosphorus. In sulphur, five different liquid phases have been observed at modest pressures (below 2 GPa). These phases are distinguished from each other by the particles that comprise the liquid being composed of rings/chains of sulphur atoms of different lengths (refs. [4.6][4.40] and refs. therein). In phosphorus, a transition line between two liquid phases has been mapped out up to 2450 K [4.41] between a phase in which the particles composing the fluid are P_4 rings and a phase in which the particles are chains of P atoms with a variety of lengths (referred to as a "polymeric phase" [4.42]).

The most well known, and probably most important, example of a transition in the second category is the transition between low density and high density H_2O. To the author's knowledge, the only other transition in the second category that has been characterized in detail is Si [4.43] though there is tentative or less detailed evidence for some other examples such as C_2H_6 [3.24]. The transition in H_2O, occurring at P,T conditions close to the lowest temperature at which it is a stable liquid, has been characterized in detail using neutron diffraction combined with empirical potential structure refinement [4.44]. Figure 4.10 shows the significantly different spatial density functions

(a) Low density water

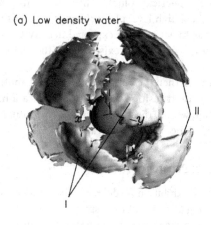

(b) High density water

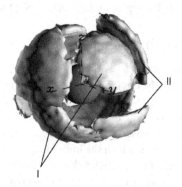

FIGURE 4.10 Spatial density functions of H_2O giving the average density around a H_2O molecule located at the origin in the y-z plane, for the low density (a) and high density (b) phases [4.44]. The labels I and II indicate locations of higher liquid density, and the difference in location of the second shell (II) is the key change between the low density and high density water phases. Reprinted figure with permission from A.K. Soper and M.A. Ricci, Phys. Rev. Lett. **84**, 2881 (2000), https://doi.org/10.1103/PhysRevLett.84.2881. Copyright (2000) by the American Physical Society.

surrounding a centrally located H_2O molecule in the y-z plane of the co-ordinate system indicated, originating from the low density and high density H_2O phases. It is somewhat unsurprising that H_2O should be a system exhibiting local structure and transitions due to the strength of the hydrogen bonding between the particles (H_2O molecules) comprising the liquid. The transition line is mapped out on the H_2O phase diagram given in Appendix B, although the extent to which the transition occurs in a sufficiently narrow P,T range to justify drawing a line on the phase diagram to represent the transition is debated.

4.4 Equations to Fit the Melting Curve on the P,T Phase Diagram

Various equations are utilized for fitting the path of the melting curve on the P,T phase diagram. The most commonly used are the Glatzel equation (equation 4.17), the Simon-Glatzel equation [4.45] (equation 4.18, essentially the same as the Glatzel equation), or the Kechin equation [4.46][4.47] (equation 4.19).

$$P = -a + bT^c \tag{4.17}$$

$$\frac{T}{T_0} = \left(1 + \frac{P - P_0}{a}\right)^{\frac{1}{b}} \tag{4.18}$$

$$\frac{T}{T_0} = \left(1 + \frac{P - P_0}{a}\right)^{\frac{1}{b}} e^{c(P - P_0)} \tag{4.19}$$

The variables a, b, c, P_0, T_0 are fitting parameters. The Glatzel and Simon-Glatzel equations have the advantage that they are invertible, the Simon-Glatzel and Kechin equations have the advantage that by definition the point P_0, T_0 always lies on the melting curve. Thus it is possible to constrain the curve to pass through a point such as the triple point instead of leaving these as free-fitting parameters. Lastly, the Kechin equation has the advantage that it can represent melting curves with negative slopes or passing through a maximum temperature.

The Simon-Glatzel equation was proposed in 1929 whilst the Kechin equation was proposed far more recently, in 1995 [4.47]. The Kechin equation is claimed to be based closely on underlying principles of thermodynamics; on the other hand, those who use the Kechin equation to fit melting curves frequently describe it as empirical. The reality is somewhere between these extremes and is best understood by describing the derivation of the Kechin equation. We do this here, following Kechin's original papers [4.46][4.47] but with notation consistent with that employed elsewhere in this text.

We begin with the Clausius-Clapeyron equation (equations 2.9 and 2.10) which we derived from first principles in Section 2.3. There we discussed the equation in relation to the vapour pressure curve, but in fact it gives the gradient of the transition/co-existence line for any first order transition so we can apply it also to the melting curve. In fact, whilst derived from the two-phase approach to melting (melting occurs when the Gibbs function of the solid and fluid phases is equal), it has also been shown to be generally consistent with the one-phase (Lindemann) approach to melting (melting occurs when the solid phase meets a certain criterion of lattice instability) [4.46]. Writing the Clausius-Clapeyron equation (equation 2.10) in terms of the volume change on melting ΔV and latent heat of melting L we obtain:

$$\frac{1}{T}\frac{dT}{dP} = \frac{\Delta V}{L} \tag{4.20}$$

We can make two statements about the melting curve straight away. Firstly since $L > 0$ always, the sign of ΔV dictates the sign of the melting curve gradient. A volume decrease upon melting must lead to $dT/dP < 0$ also, as is observed for H_2O. Secondly, if there is a triple point between two solid phases (let's call them A and B) separated by a first order phase transition and the fluid phase then there will most likely be a discontinuous change in the gradient of the melting curve. If we melt solid phase A or B at the triple point we will obtain a fluid phase with the same volume yet have started out with different solid volumes. Therefore ΔV necessarily undergoes a discontinuous change at this triple point and so does dT/dP (L is also likely to change).

We would like to find a way to integrate this equation so that we can obtain a function $T(P)$ describing the path of the melting curve. For most real fluids, the beginning of the melting curve at the triple point has been measured with extremely high accuracy and precision and melting data are available up to high pressures and temperatures with progressively larger errors due to the challenges of making measurements at such extreme conditions. Therefore, an appropriate approach is to perform a Taylor expansion(s) on the parameter $\Delta V/L$ about a fixed reference point P_0, T_0 which is often (but not necessarily) the triple point. The greatest flexibility in the form of the final equation for the melting curve would be provided by a single infinite Taylor series. The greatest flexibility offered by a truncated Taylor series is given by expressing $\Delta V/L$ as the ratio of two truncated Taylor series:

$$y = \frac{\sum_{j=0}^{L} c_j x^j}{\sum_{k=0}^{M} b_k x^k} \tag{4.21}$$

Where we have used the notations:

$$x = P - P_0$$

$$y = \frac{\Delta V}{L}$$

We can choose values for L, M dependent on the degree of precision we require and the degree of complexity we are prepared to accept. For all L, M we obtain:

$$\left(\frac{dT}{dP}\right)_{P=P_0} = \frac{T_0 c_0}{b_0} \tag{4.22}$$

This expression may be used to constrain the fitting parameters if $\Delta V, L$ and/or dT/dP have been measured accurately at the reference point P_0, T_0.

Integrating equation 4.20 to obtain $T(P)$ for the melting curve, we obtain:

$$\int_{T_0}^{T} \frac{dT}{T} = \int_{x=0}^{P-P_0} y\,dx$$

$$\ln\left(\frac{T}{T_0}\right) = F_{LM}(x)$$

Where:

$$F_{LM}(x) = \int_{x=0}^{x=P-P_0} \frac{\sum_{j=0}^{L} c_j x^j}{\sum_{k=0}^{M} b_k x^k}\,dx \tag{4.23}$$

Performing this integration, we obtain the Kechin equation for $F_{11}(x)$ and the Simon-Glatzel equation for $F_{01}(x)$. In the notation of equation 4.21, the Simon-Glatzel equation is:

$$\frac{T}{T_0} = \left[1 + \frac{b_1}{b_0}(P - P_0)\right]^{\frac{c_0}{b_1}}$$

There are effectively two parameters to fit, b_1 / b_0 and c_0 / b_1, and the fit passes through P_0, T_0 regardless of their values. It is therefore possible to also constrain the gradient at P_0, T_0 using equation 4.22 if this is known accurately enough to justify doing so.

The parameters of the Kechin equation as given earlier (equation 4.19) are, in terms of c_j, b_k:

$$a = \frac{b_0}{b_1}$$

$$\frac{1}{b} = \frac{c_0}{b_1} - \frac{c_1 b_0}{b_1^2}$$

$$c = \frac{c_1}{b_1}$$

Thus, in this case also we can constrain the gradient at P_0, T_0 using equation 4.22 if we wish to do so.

The Kechin equation can, equivalently, be derived from a single truncated Taylor series to which the Padé approximation has been applied [4.47]. The approximation in the Padé method is no less valid than simply making a direct truncation of the Taylor series.

The original papers by Kechin [4.46][4.47] discuss constraints on c_j, b_k to ensure that $d^2T / dP^2 < 0$ for all x, i.e., the melting curve cannot exhibit concavity towards the temperature axis on the linear P, T phase diagram under any circumstances. This is justified on the basis that this is not observed in experimentally measured melting curves, but a lack of concavity towards the temperature axis cannot be justified for all fluids from first principles. The only melting curve the author is aware of that exhibits concavity towards the temperature axis that cannot be explained by experimental error or a solid-solid phase transition meeting the melting curve is Cerium [4.6].

When the Kechin or Simon-Glatzel equation is fitted to experimental data it should not exhibit concavity towards the temperature axis within the range of the data, provided that the data do not exhibit this feature. Whilst the Kechin and Simon-Glatzel equations should generally produce physically reasonable results when extrapolated beyond the range of data used for the fit, it is not reasonable to expect that this will *always* be the case. This is because, as we have seen, they both originate from truncated Taylor series which are only correct in the low x limit and cannot be expected to behave correctly at all times when x is large. The condition $c_0^2 - b_1 c_0 < 0$ constrains the Simon-Glatzel equation to not show concavity towards the temperature axis at any x and the condition $c_0^2 + c_1 b_0 - b_1 c_0 < 0$ constrains the Kechin equation to not show concavity towards the temperature axis for $x = 0$.

The Kechin equation can reproduce a maximum in the melting curve, but not a minimum. Neither of these equations can reproduce a kink, or discontinuity in the gradient of, a melting curve. Thus, if a triple point exists between the fluid phase and a first order transition between two solid phases it is necessary to utilize separate equations to describe the melting curve on each side of this triple point.

As we will see in Appendices A and B the Kechin and Simon-Glatzel equations are very widely used to fit to experimentally measured melting curves, yet there is no simple answer to the question posed at the beginning of this section: To what extent are they based on first principles? By definition, a Taylor series becomes a less reasonable approximation of the function it represents as the value of x increases. In addition, we have seen that the gradient of the melting curve is determined exactly by quantities which are, in principle, experimentally measurable ($\Delta V, L$). Thus, if the Kechin equation is used to fit a melting curve close to a known reference point P_0, T_0 and the curve is constrained so that the gradient is in agreement with some known values of $\Delta V, L$ then the equation for the melting curve is based closely on first principles. In practice the Simon-Glatzel and Kechin equations are utilized to fit melting curves at extreme P, T obtained in DAC experiments. In this case $P \gg P_0$, the error on ΔV is large and L cannot be measured. In this case it is reasonable to describe both equations as empirical.

4.5 What Happens to the Melting Curve in the High P, T Limit?

The vapour pressure curve ends at the critical point. On the other hand, it is generally believed that the melting curve does not end at a critical point, due to the qualitative difference between the solid and liquid states. But this has not stopped people wondering about it over the years [2.8][4.48]. In *Statistical Physics*, Mandl outlines his belief that the melting curve does not end at a critical point, before admitting that his arguments are "not accepted by everyone" [2.1]. The argument against the melting curve ending at a critical point is based particularly on the fundamental difference between the solid and liquid states in the level of static order and symmetry. The symmetry of a crystalline solid is broken by the lattice and basis vectors having specific directions whilst both fluids and gases are completely isotropic over long distances.

However, it is tempting to revisit the critical point argument in light of the modern proposal (outlined elsewhere in this text) that there is no fundamental difference between the solid and liquid states. Liquids can display a surprisingly high degree of structural order, whilst solids can display a surprisingly low degree of structural order.

For the melting curve to end at a critical point, the latent heat and volume change would both have to tend towards zero on P,T increase. In this section we review a representative sample of the experimental data showing that there is little, if any, sign of this happening. Similar data for other fluids and/or information on alternate mechanisms to terminate the melting curve are shown in the appendices. Also in this section, we will outline the other factors that can prevent the melting curve from reaching the high P,T limit. It is therefore unlikely that a melting curve ending at a critical point will ever be observed.

For all fluids, there are mechanisms to prevent the melting curve continuing to arbitrarily high P,T. In a molecular fluid, at a certain point the temperature will become high enough to begin breaking the interatomic bonds. We discussed in Chapter 1 how this begins at ca. 3000 K for H_2. Even in an atomic fluid, at some point (ca. 10^5 K for Ar) there will be enough thermal energy to remove the weakest bound electrons from the atoms, fundamentally altering the nature of the fluid. Further temperature increase will remove more electrons until eventually the fluid transforms entirely into a plasma.

Separately from this it can be argued that, for quantum mechanical reasons, melting curves can eventually reach a temperature maximum, then decrease in temperature upon further pressure increase. Detailed discussion of this phenomenon is beyond the scope of this textbook, but we will outline the basics here. As pressure is increased on a sample, regardless of its state or structure, the diameter Δx of the "box" occupied by each atom decreases. Thus the position of the atom becomes constrained to a smaller space and its momentum must therefore increase due to the uncertainty principle $\Delta x \Delta p \geq \hbar/2$. Eventually, this momentum/zero-point energy prevents the sample from having a stable static structure to put it in the solid state. H_2 is the principal melting curve for which detailed theoretical calculations unequivocally support this hypothesis, for other melting curves it is still very much an open question [4.49].

Presently, observing this phenomenon experimentally requires pressures at the very limit of what is possible experimentally and negative gradient melting curves due to this phenomenon have only been observed in the lightest elements. Essentially, the problem is as follows. Shock compression generates far too high a temperature to study the effects of zero-point energy in a meaningful way. Static compression in the DAC cannot reach the required pressures, exacerbated by the fact that the lightest elements H_2, He, and Li easily escape from the pressure chamber by diffusing through the diamonds and/or gasket. Further discussion is therefore beyond the scope of this text, and readers are referred to other literature on the subject [4.49][4.50][4.51][4.52].

Even studying the melting curves of fluids comprised of particles heavy enough for the above quantum mechanical arguments not to apply at accessible P,T, the experimental data gathered to date does not support the hypothesis of the melting curve ending at a critical point or exhibiting other anomalous behaviour at extreme P,T. We examine the examples of Ar and N_2.

The melting curve of Ar has been studied using a variety of methods, but the highest P,T measurements were those of Abramson[4.53] in the DAC. These measurements extended up to 6 GPa, 750 K. The $P(T)$ melting points were combined with previous studies and fitted using the Simon-Glatzel equation (equation 4.18 and Figure 4.11 (a)). The gradient of the melting line on the P,T phase diagram does decrease slightly at high P,T but certainly not to the extent that it constitutes evidence for getting close to a maximum and temperature decrease upon further pressure increase.

Measurements of the density change upon melting have been made up to ca. 300 K (ref. [4.2] and refs. therein). The melting curve data begin at a temperature close to the triple point and initially the density change decreases upon temperature increase, before settling down to—considering the spread in the data—a constant value (Figure 4.11 (b)). Certainly, there is no evidence of a trend towards zero and the Ar melting curve trending towards a temperature maximum or termination at a critical point.

We see a similar picture with N_2. Figure 4.12 (a) shows the melting curve on the P,T phase diagram from the triple point up to the highest temperature at which the fit is backed by experimental data, from ref. [4.5] and refs. therein. The melting temperature continues to shift almost linearly upwards upon pressure increase (fitted using the Glatzel equation, equation 4.17). Figure 4.12 (b) shows the volume change ΔV upon melting for the same temperature range. ΔV continues to decrease up to the highest temperature studied, but the data and fit do not indicate that $\Delta V \rightarrow 0$ is likely upon further temperature increase.

4.6 Summary

In this chapter we have examined the static properties of fluids confined under P,T conditions close to the melting curve. We commenced by wishing to understand the liquid state close to the melting curve but have looked at many examples of fluids close to the melting curve in which $T > T_C$ by a significant amount. We shall see in Chapter 6 that it is correct to describe these fluids, at significantly supercritical density, as liquids—despite $T > T_C$. At density close to the melting curve we have seen that the static properties of these fluids (particularly the fundamental properties of density and compressibility) are close to those of solids. The increase in density upon crossing the melting curve is tiny and is even, for the important example of H_2O, negative. This is a marked contrast to the vapour pressure curve.

As a result of the solid-like density when the melting curve is approached, fluids under these conditions can exhibit properties such as short-range order that we would traditionally associate only with solids. Experimentally, the solid-like density results in the possibility to obtain diffraction data good enough to warrant detailed analysis using Fourier transform methods or empirical potential structure refinement. A good deal of humankind's knowledge of the properties of dense fluids is based on diffraction measurements.

From the examples shown in the previous section and in the appendices we can draw two important conclusions regarding the melting curve itself. Firstly, with the exception of fluids composed of extremely light particles H_2, He, etc., any maximum that may exist in the melting curve does not lie at currently experimentally accessible P,T.

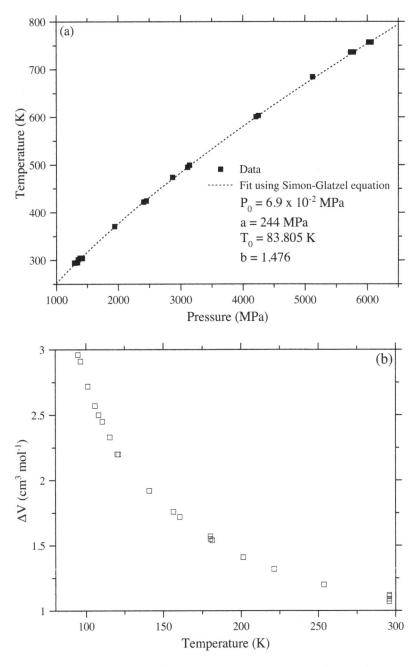

FIGURE 4.11 (a) Ar melting points measured experimentally and fitted using the Simon-Glatzel equation. Adapted from ref. [4.53] with permission. (b) Density change upon melting from the triple point up to $2.1T_C$ (data obtained from refs. within ref. [4.2]).

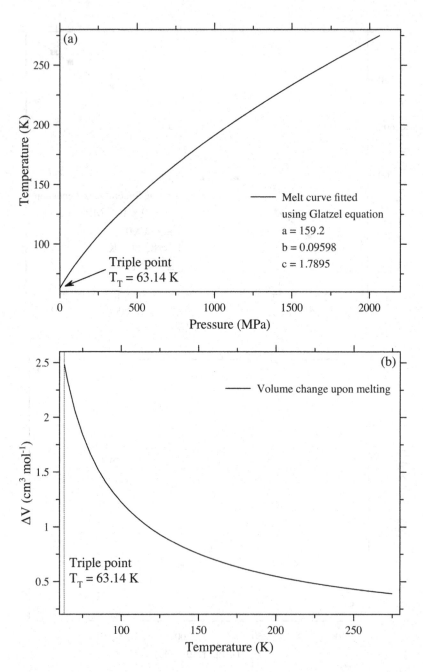

FIGURE 4.12   (a) Glatzel equation fit to melting curve of N_2 on the P,T phase diagram. (b) Fit to the volume change upon melting as a function of temperature. Fits are displayed from the beginning of the melting curve at the triple point, up to the highest temperature at which the fits are backed by experimental data. Data from ref. [4.5] and refs. therein.

Secondly, the evidence does not support the hypothesis that the melting curve may ever end at a critical point. For this we would need to observe parameters such as the volume change upon melting and latent heat of melting trending towards zero upon temperature increase. Whilst these parameters do decrease upon increase in temperature, they are not trending towards zero.

References

4.1 Ch. Tegeler, R. Span and W. Wagner, J. Phys. Chem. Ref. Data **28**, 779 (1999).

4.2 R.A. Wilsak and G. Thodos, J. Chem. Eng. Data **29**, 255 (1984).

4.3 R. Span, E.W. Lemmon, R.T. Jacobsen, W. Wagner and A. Yokozeki, J. Phys. Chem. Ref. Data **29**, 1361 (2000).

4.4 V.M. Cheng, W.B. Daniels and R.K. Crawford, Phys. Rev. B **11**, 3972 (1975).

4.5 T.A. Scott, Phys. Rep. **27**, 85 (1976).

4.6 , E.Yu. Tonkov and E.G. Ponyatovsky, *phase transformations of elements under high pressure*, CRC Press, Boca Raton (2005).

4.7 W.E. Streib, T.H. Jordan and W.N. Lipscomb, J. Chem. Phys. **37**, 2962 (1962).

4.8 J.E. Proctor, C.G. Pruteanu, I. Morrison, I.F. Crowe and J.S. Loveday, J. Phys. Chem. Lett. **10**, 6584 (2019).

4.9 F.D. Murnaghan, Am. J. Math. **59**, 235 (1937).

4.10 W.B. Holzapfel, Equations of state for solids, in *High pressure physics*, J.S. Loveday (ed.), CRC Press, Boca Raton (2012).

4.11 https://www.isis.stfc.ac.uk/Pages/Gudrun.aspx and references therein.

4.12 J.H. Eggert, G. Weck, P. Loubeyre and M. Mezouar, Phys. Rev. B **65**, 174105 (2002).

4.13 C. Prescher, Yu.D. Fomin, V.B. Prakapenka, J. Stefanski, K. Trachenko and V.V. Brazhkin, Phys. Rev. B **95**, 134114 (2017).

4.14 A.K. Soper, Mol. Sim. **38**, 1171 (2012).

4.15 https://www.isis.stfc.ac.uk/Pages/Empirical-Potential-Structure-Refinement.aspx

4.16 F. Hajdu, Acta Cryst. A **28**, 250 (1972).

4.17 D. Smith et al., Phys. Rev. E **96**, 052113 (2017).

4.18 J.E. Proctor, D.A. Melendrez Armada and A. Vijayaraghavan, *An introduction to graphene and carbon nanotubes*, CRC Press, Boca Raton (2017).

4.19 G.E. Ice, J.D. Budai and J.W.L. Pang, Science **334**, 1234 (2011).

4.20 B.E. Warren, *X-ray diffraction*, Addison-Wesley (1969).

4.21 A.K. Soper and E.R. Barney, J. Appl. Cryst. **44**, 714 (2011).

4.22 H.E. Fischer, A.C. Barnes and P.S. Salmon, Rep. Prog. Phys. **69**, 233 (2006).

4.23 S. Kohara et al., J. Phys.: Cond. Mat. **19**, 506101 (2007).

4.24 U. Hoppe, G. Walter, R. Kranold and D. Stachel, J. Non-Cryst. Sol. **263 & 264**, 29 (2000).

4.25 C. Sanloup, E. Gregoryanz, O. Degtyareva and M. Hanfland, Phys. Rev. Lett. **100**, 075701 (2008).

4.26 V.M. Giordano, F. Datchi and A. Dewaele, J. Chem. Phys. **125**, 054504 (2006).

4.27 E. Lorch, J. Phys. C (solid state physics) **2**, 229 (1969).

4.28 H.E. Fischer, A.C. Barnes and P.S. Salmon, Rep. Prog. Phys. **69**, 233 (2006).

4.29 A.C. Wright, J. Non-cryst. Sol. **179**, 84 (1994).

4.30 L.B. Skinner, C. Huang, D. Schlesinger, L.G.M. Petterson, A. Nilsson and C.J. Benmore, J. Chem. Phys. **138**, 074506 (2013).

4.31 G. Etherington, A.C. Wright, J.T. Wenzel, J.C. Dore, J.H. Clarke and R.N. Sinclair, J. Non-cryst. Sol. **48**, 265 (1982).

4.32 K. Laaziri et al., Phys. Rev. B **60**, 13520 (1999).

4.33 J.B. Bates, J. Chem. Phys. **61**, 4163 (1974).

4.34 M. Micoulaut, L. Cormier and G.S. Henderson, J. Phys.: Cond. Mat. **18**, R753 (2006).

4.35 R.L. McGreevy and L. Pusztai, Mol. Sim **1**, 359 (1988).

4.36 A. Baranyai and D.J. Evans, Phys. Rev. A **40**, 3817 (1989).

4.37 A.K. Soper, J. Chem. Phys. **150**, 234503 (2019).

4.38 G. Weck et al., Phys. Rev. B **80**, 180202 (2009).

4.39 V.V. Brazhkin and A.G. Lyapin, J. Phys.: Cond. Mat. **15**, 6059 (2003).

4.40 *Polymorphism of sulfur: Structural and dynamical aspects*. Ph.D. thesis of Laura Crapanzano, Université Joseph-Fourier (2006). https://tel.archives-ouvertes.fr/tel-00204149

4.41 G. Monaco, S. Falconi, W.A. Crichton and M. Mezouar, Phys. Rev. Lett. **90**, 255701 (2003).

4.42 D. Hohl and R.O. Jones, Phys. Rev. B **45**, 8995 (1992).

4.43 S. Sastry and C.A. Angell, Nat. Mater. **2**, 739 (2003).

4.44 A.K. Soper and M.A. Ricci, Phys. Rev. Lett. **84**, 2881 (2000).

4.45 F. Simon and G. Glatzel, Z. Anorg. Allg. Chem. **178**, 309 (1929) (in German).

4.46 V.V. Kechin, Phys. Rev. B **65**, 052102 (2001).

4.47 V.V. Kechin, J. Phys.: Cond. Mat. **7**, 531 (1995).

4.48 A. Aitta, J. Stat. Mech.: Theory and Experiment **2006**, 12015 (2006).

4.49 C.L. Guillaume et al., Nat. Phys. **7**, 211 (2011).

4.50 J.M. McMahon, M.A. Morales, C. Pierleoni and D.M. Ceperley, Rev. Mod. Phys. **84**, 1607 (2012).

4.51 J. Chen et al., Nat. Comm. **4**, 2064 (2013).

4.52 N.W. Ashcroft, J. Phys.: Cond. Mat. **12**, A129 (2000).

4.53 E.H. Abramson, High Press. Res. **31**, 549 (2011).

5

The Liquid State Close to the Melting Curve (II): Dynamic Properties

In this chapter we will study the dynamic properties of the liquid state close to the melting curve. We will focus on the Maxwell and Frenkel models permitting the liquid to exhibit viscous (liquid-like) and elastic (solid-like) responses to stress on different timescales and advancements in the past decade in our ability to predict—arguably—the most important fluid dynamic property (heat capacity) using the Frenkel model. We will also describe the experimental methods utilized to study the variety of excited states existing in liquids: Raman and Brillouin spectroscopy and inelastic neutron and X-ray scattering. From use of these methods we have, in some cases, quite detailed knowledge of the sound wave propagation in liquids, vibrational and rotational excited states in molecular liquids, etc.

5.1 Phonon Theory of Liquids

The phonon theory of liquids is the approach to modelling the dynamic properties of liquids that begins from the assumption that their behaviour is solid-like and incorporates their ability to flow as a correction to this; building on the ideas of Maxwell and Frenkel. We will begin with the Maxwell model and then move on to the Frenkel model which has qualitative similarities to Maxwell but is atomistic so can be considered in relation to properties such as the heat capacities and velocity autocorrelation function (VAF) that relate to the fluid behaviour on a microscopic scale.

Both models incorporate the liquid's ability to exhibit an elastic response to applied stress on a short timescale, then to flow on a longer timescale. Thus they serve to demarcate the timescales on which the liquid exhibits solid-like behaviour (elastic response) and gas-like behaviour (the ability to flow).

5.1.1 Frenkel and Maxwell Models

The simplest form of the Maxwell model is a mathematical representation of a spring and dashpot in series (Figure 5.1). The applied force F and extension x in the system are

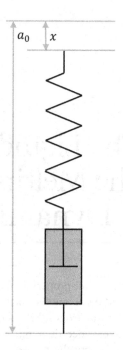

FIGURE 5.1 Spring and dashpot model of the viscous and elastic responses of a fluid to strain. An applied strain (displacement x from the equilibrium length a_0) will initially be absorbed elastically by the spring, then gradually transferred to the dashpot containing a viscous fluid.

linked by equation 5.1. The spring is described by the spring constant k and the dashpot by the viscosity η.

$$\frac{dx}{dt} = \frac{1}{k}\frac{dF}{dt} + \frac{F}{\eta}$$ (5.1)

The two terms on the right-hand side of this equation describe the elastic and viscous parts of the system's response respectively. Thus the entire response is described as *viscoelastic*. When we (for instance) instantaneously compress the system by a given amount (x as shown in Figure 5.1) the compression is initially taken entirely by the spring, which then gradually extends back as the compression is transferred to the dashpot. For this case (constant x) we obtain:

$$F = F_0 e^{-\frac{kt}{\eta}}$$

We can see that the force required to maintain the compression decays with a time constant (the Maxwell relaxation time) depending on the spring constant and viscosity: $\tau = \eta / k$. Alternatively [1.14], the model may be applied to the case of a shear stress in a three-dimensional system in which case the shear modulus G takes the place of the spring constant k and we obtain equation 5.2.

$$\tau_M = \frac{\eta}{G} \tag{5.2}$$

Whilst the Maxwell relaxation time has been used in a semi-empirical manner to model shear wave propagation in fluids [1.11], the key contribution made by Frenkel [5.1] was to develop the concept of liquid relaxation time on an atomistic level. In the liquid state close to the melting curve we can conclude, from consideration of the density alone, that the particles spend most of their time constrained to a certain equilibrium position similarly to atoms in a solid. This condition is met for most of the liquids we may encounter in everyday life—water, cooking oil, acetone, liquid nitrogen or argon in the laboratory, etc. Therefore, the motion of the particles most of the time is vibration about a fixed equilibrium position the same as in a solid. Only occasionally does a particle make a jump to swap places with an adjacent particle. It is this feature, however, which gives the liquid its ability to flow. The liquid relaxation time as defined by Frenkel (τ_R) is the average time spent by an individual particle in an equilibrium position between successive jumps. Whilst the Maxwell and Frenkel relaxation times are defined in different ways (macroscopic vs. microscopic), they are the same order of magnitude [1.14][5.1]. The sequence of schematics in Figure 5.2 demonstrates the oscillatory motion of particles in a fluid around an equilibrium position followed by a jump of the marked particle to a new equilibrium position.

Most readers will be familiar with the fact that liquids cannot support all shear waves. The seismic s-waves that cannot penetrate the earth's liquid outer core are a good

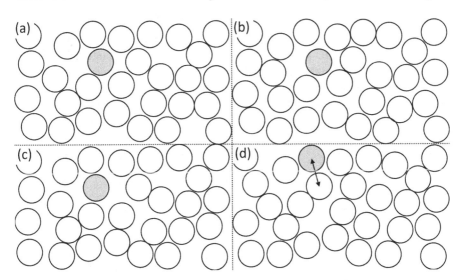

FIGURE 5.2 Sequence of schematic images showing fluid particles undergoing oscillatory motion about equilibrium positions followed by the shaded particle making a jump to exchange places with an adjacent particle.

example of shear waves that the liquid in question cannot support. However, what is less well known is that most liquids, under most conditions, can support *some* shear waves. Frenkel's atomistic model of the dense fluid outlined above permits this. Approximately speaking, shear waves can be supported provided that they have a period smaller than τ_R. That is, that the particles stay in their equilibrium positions long enough for the shear wave to propagate. This sets the minimum frequency and energy of the shear waves according to the inequality 5.3.

$$\omega \gtrsim \frac{2\pi}{\tau_R} \qquad (5.3)$$

The inequality here is approximate. Nonetheless as we see in the next section, it can be incorporated into a method to successfully model the heat capacities of many liquids (and supercritical fluids) over a wide P,T range.

5.1.2 Prediction of Liquid Heat Capacity

The phonon[*] theory of liquid thermodynamics [1.14][1.15] predicts the heat capacity of liquids using the model above to account for their ability to support shear waves in some cases. Whilst the effect on heat capacity is the most important consequence of the liquid's ability to support shear waves, it is important to note that the propagation of shear waves in liquids has also been observed directly by many research groups over the years with a variety of different experimental methods—for instance frequency-domain Brillouin scattering since the 1960s as discussed in Section 5.3, time-domain Brillouin scattering [5.2], inelastic X-ray scattering [5.3] and studies of ultrasound propagation [5.4].

To describe liquid and supercritical fluid heat capacity using this method (as outlined in ref. [2.1]), we can begin with the expressions from Landau and Lifshitz [1.4] for the Helmholtz function F of a solid and the link between this and the internal energy U (equation 5.4). The subscripts ph denote that we are considering at present only the contribution from solid-like phonons. The expressions consider the solid with vibrational degrees of freedom, with frequencies ω_i. The zero-point energy term has been omitted[†].

$$F_{ph} = k_B T \sum_i \ln\left[1 - e^{-\frac{\hbar\omega_i}{k_B T}} \right]$$

$$U_{ph} = F_{ph} - T \frac{dF_{ph}}{dT} \qquad (5.4)$$

---

[*] The term "phonon" is used since we are tackling the problem using the theoretical methods used to characterize phonons in solids. A shear wave propagating in a fluid due to satisfying the condition $\omega \gtrsim \frac{2\pi}{\tau_R}$ is analogous to a shear phonon propagating in a solid.

[†] All fluids except He solidify upon temperature decrease before zero-point energy becomes relevant, and also the physically important parameter of heat capacity is obtained by differentiating the expression for the energy.

The heat capacity per particle is obtained from the internal energy in the usual manner:

$$c_V = \frac{1}{N}\left(\frac{\partial U}{\partial T}\right)_V \tag{5.5}$$

We need to account for the fact that the vibrational frequencies ω_i are temperature dependent. For simplicity let's begin with the case of a dense noble fluid which is isotropic and described by the Lennard-Jones potential. In this case ω_i is linked to the volume and therefore also the thermal expansion coefficient as follows in the solid-like approach [5.5]:

$$\frac{\omega_i}{\omega_i^0} = \left[\frac{V(T)}{V_0}\right]^{-\gamma}$$

In many materials γ (the Grüneisen parameter) varies considerably between different vibrational modes, but typically takes values in the range 1–5. We show later (Section 5.2.1) that for a potential with the Lennard-Jones (6-12) form $\gamma = 3.5$ exactly; $\gamma = 0.5$ has also been used [1.15], and lower values can be observed at really extreme conditions (solids at 10–100 GPa pressures). In this case we obtain equation 5.6, where the volume coefficient of thermal expansion β is defined in the usual manner‡.

$$\left(\frac{\partial \omega}{\partial T}\right)_P = -\gamma\omega\beta \tag{5.6}$$

Using this, we obtain the following expression for U_{ph}:

$$U_{ph} = \left(1+\gamma\beta T\right)\sum_i \frac{\hbar\omega_i}{e^{\frac{\hbar\omega_i}{k_B T}}-1} \tag{5.7}$$

All that remains now is to evaluate the contributions to U from the longitudinal and transverse modes respectively, accounting for the fact that the transverse (shear) modes with frequencies below ω_F cannot propagate. Let's begin with the longitudinal modes. The sum in equation 5.7 becomes an integral over all frequencies up to the Debye frequency ω_D where the density of modes as a function of frequency is given by the density of states $g_L(\omega)$. This is normalized so that there are N modes in total up to ω_D. This gives us $g_L(\omega) = 3N\omega^2 / \omega_D^3$. The internal energy due to the longitudinal modes U_L is:

$$U_L = \left(1+\gamma\beta T\right)\int_0^{\omega_D} \frac{3N}{\omega_D^3} \times \frac{\hbar\omega^3 d\omega}{e^{\frac{\hbar\omega}{k_B T}}-1}$$

---

‡ $\beta = \dfrac{1}{V}\left(\dfrac{\partial V}{\partial T}\right)_P$

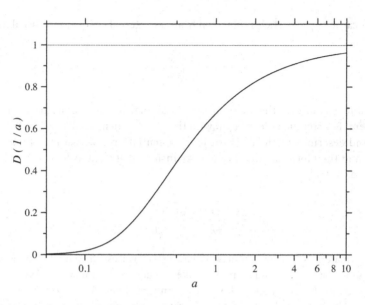

FIGURE 5.3 Plot of the Debye function $D\left(\frac{1}{a}\right)$ as a function of a, equivalent to the effect of a temperature increase in the terms appearing in equation 5.9.

This can be evaluated to give:

$$U_L = \left(1 + \gamma \beta T\right) \times N k_B T D\left(\frac{\hbar \omega_D}{k_B T}\right)$$

Where $D(x)$ is the Debye function (equation 5.8, where we integrate over values of z up to x, and $x = \hbar \omega_D / k_B T$ in this case). Figure 5.3 plots the Debye function $D(1/a)$ as this is analogous to a plot against increasing temperature. The Debye function saturates at 1 when there is ample thermal energy available to excite modes up to the Debye frequency ($\hbar \omega_D \ll k_B T$).

$$D(x) = \frac{3}{x^3} \int_0^x \frac{z^3 dz}{e^z - 1} \qquad (5.8)$$

We perform an equivalent procedure to obtain the internal energy due to transverse modes, except that this time there are $2N$ modes from $\omega = 0$ to ω_D but only those modes with $\omega > \omega_F$ can be excited and we use the notation $U_T\left(\omega > \omega_F\right)$. We obtain:

$$U_T\left(\omega > \omega_F\right) = 2\left(1 + \gamma \beta T\right) \int_{\omega_F}^{\omega_D} \frac{3N}{\omega_D^3} \times \frac{\hbar \omega^3 d\omega}{e^{\frac{\hbar \omega}{k_B T}} - 1}$$

$$U_T\left(\omega > \omega_F\right) = 2 N k_B T\left(1 + \gamma \beta T\right)\left[D\left(\frac{\hbar \omega_D}{k_B T}\right) - \left(\frac{\omega_F}{\omega_D}\right)^3 D\left(\frac{\hbar \omega_F}{k_B T}\right)\right] \qquad (5.9)$$

However, the total internal energy of the sample also contains contributions from diffusive motion of the particles, which we have not evaluated. This problem is solved using the virial theorem [1.15]. The total internal energy U of the sample can be expressed as the kinetic (K) and potential (P) energies due to longitudinal vibrational modes (L), transverse vibrational modes (T), and diffusive motion (D).

$$U = K_L + P_L + K_T(\omega > \omega_F) + P_T(\omega > \omega_F) + K_D + P_D$$

The term P_D is negligible because the restoring forces at low frequency are extremely weak—otherwise transverse modes at low frequency would exist and gas-like diffusion would not be possible. Discarding P_D and grouping the relevant terms into the total kinetic energy K of the sample we obtain:

$$U = K + P_L + P_T(\omega > \omega_F)$$

Using the virial theorem we can write:

$$P_L = \frac{U_L}{2}$$

$$P_T(\omega > \omega_F) = \frac{U_T(\omega > \omega_F)}{2}$$

$$K = \frac{U_L + U_T}{2} = \frac{U_L + U_T(\omega > \omega_F) + U_T(\omega < \omega_F)}{2}$$

Using these expressions we obtain:

$$U = U_L + U_T(\omega > \omega_F) + \frac{U_T(\omega < \omega_F)}{2} \tag{5.10}$$

We therefore need to evaluate the (hypothetical) internal energy due to shear waves with $\omega < \omega_F$. We obtain using the same methodology as earlier:

$$U_T(\omega < \omega_F) = 2(1+\gamma\beta T)\int_0^{\omega_F} \frac{3N}{\omega_D^3} \times \frac{\hbar\omega^3 d\omega}{e^{\frac{\hbar\omega}{k_B T}} - 1} = 2(1+\gamma\beta T)Nk_BT\left(\frac{\omega_F}{\omega_D}\right)^3 D\left(\frac{\hbar\omega_F}{k_B T}\right)$$

Equation 5.10 then yields:

$$U = Nk_BT(1+\gamma\beta T)\left[3D\left(\frac{\hbar\omega_D}{k_B T}\right) - \left(\frac{\omega_F}{\omega_D}\right)^3 D\left(\frac{\hbar\omega_F}{k_B T}\right)\right] \tag{5.11}$$

This final expression for U may then be differentiated numerically to obtain the heat capacity per particle (c_V) or per mole (C_V) by substituting $N = N_A$, $R = N_A k_B$. It has so far been used to model the heat capacities of a wide variety of fluids, albeit over a limited

P,T range [1.15]. In this modelling, all parameters which are experimentally measured take their measured values and all parameters which cannot be measured take physically reasonable values obtained via trial and error. We will work through two examples here (Hg at ambient pressure and Ar at 378 MPa).

For all fluids, it is necessary to begin by estimating the Frenkel frequency as a function of temperature. This is done by equating it to the frequency calculated from the Maxwell relaxation time given in equation 5.2. These considerations result in equation 5.12 for ω_F:

$$\omega_F(T) = \frac{2\pi G_\infty}{\eta(T)} \qquad (5.12)$$

Experimental data are available in the literature for the viscosity $\eta(T)$ of liquid Hg [5.6] (Figure 5.4). To calculate C_V it is necessary to use equation 5.11 to calculate U at closely spaced values of T, allowing numerical differentiation to obtain C_V. We therefore fit a function to $\eta(T)$ so that we can interpolate between the experimental measurements:

$$\eta(T) = 0.612 \times e^{235.175/(T-40.276)} \qquad (5.13)$$

The parameter G_∞ in equation 5.12 is the shear modulus in the infinite frequency limit. Experimental measurements of this parameter are very rare, but the value given of 1.31 GPa is consistent with the limited data available [1.15]. We additionally require the

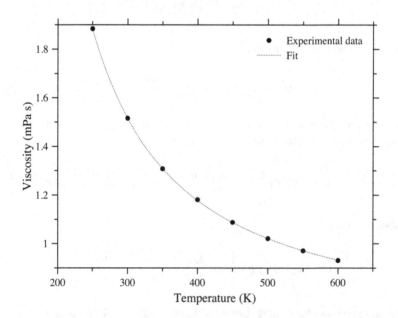

FIGURE 5.4 Experimental data for the viscosity of liquid Hg from melting to 600 K [5.6] and fit using equation 5.13.

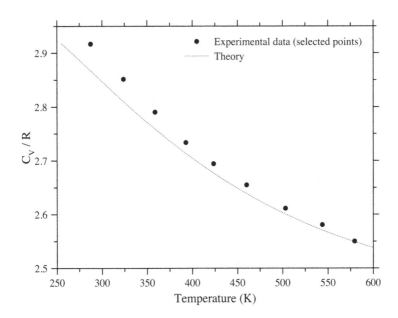

FIGURE 5.5 Predicted trend in heat capacity C_V for liquid Hg compared to selected experimental data points following subtraction of the electronic contribution [5.7].

value for the Debye frequency ω_D. This also cannot be measured experimentally but is given the physically reasonable value of 1.28×10^{13} rad s^{-1}. The thermal expansion coefficient β appearing in equation 5.11 can be experimentally measured and is temperature dependent. It is, however, multiplied by the Grüneisen parameter γ which is known only roughly. We therefore use the experimentally measured average value of β over the temperature range considered here ($\beta = 1.4 \times 10^{-4}$ K^{-1}) and the physically reasonable value of $\gamma = 0.45$. Using these values we obtain the trend in C_V shown in Figure 5.5 and compared to experimental measurements. The experimental data have had the electronic contribution subtracted [5.7].

To produce the data in Figure 5.5 we have followed fairly closely the methodology of ref. [1.15] and obtained similar results. This can also be applied to noble gas fluids and to molecular fluids (provided the contribution to C_V from intramolecular degrees of freedom can be subtracted accurately). In Figure 5.6 we show experimental and calculated C_V for fluid Ar at 378 MPa. This has been calculated in the harmonic approximation (no thermal expansion) using parameters $\omega_D = 1.26 \times 10^{13}$ rad s^{-1} and $G_\infty = 0.2$ GPa. For Ar and other noble gas fluids accurate values for the viscosity can be obtained from the fundamental EOS via NIST REFPROP so it is not necessary to perform an interpolation between a limited set of datapoints as we did for Hg. On the other hand, the calculation using equation 5.11 can provide a C_V matching far more closely the experimental values from NIST REFPROP if a fit to the viscosity values using the VFT (Vogel-Fulcher-Tammann) equation is utilized in the calculation of ω_F (equation 5.12) instead of the raw output from the fundamental EOS.

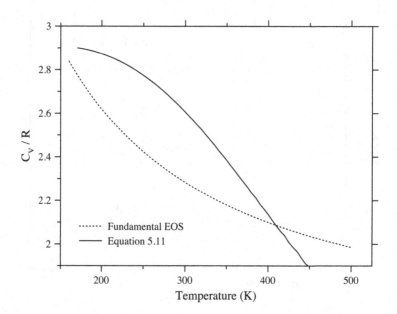

FIGURE 5.6 Temperature dependence of C_V of fluid Ar at 378 MPa obtained from the fundamental EOS via NIST REFPROP (dotted line) and using the phonon theory of liquid thermodynamics as outlined in the text and equation 5.11 (solid line).

5.2 Raman Spectroscopy of Liquids and Supercritical Fluids Close to the Melting Curve

In Chapter 1 (Section 1.4) we reviewed the vibrational Raman spectra of gases; namely that the frequency (energy) shifts downwards upon isothermal pressure increase and upwards upon isobaric temperature increase. The same behaviour is generally observed in gas-like liquids close to the vapour pressure curve, for instance in C_2H_6 (Figure 5.7) [3.24] and CO_2 [5.8][5.9]. In both cases, and others, this behaviour is observed at subcritical temperature on the liquid side of the vapour pressure curve as well as at supercritical temperature at gas-like density. Just as in the cases of gases discussed in Chapter 1, the principal effect of an isothermal pressure increase is to increase the influence of the attractive van der Waals forces between molecules, thereby loosening the intramolecular bond rather than forcing the atoms in the individual molecule closer together.

In this section we consider the Raman spectra of liquids and supercritical fluids close to the melting curve. Whilst most of the data available in the literature are on Raman spectra, absorption/emission spectroscopy is also possible. In this region of the phase diagram the evolution of the vibrational Raman spectra upon P,T change usually observed is the exact opposite to that observed in gas-like liquids. Instead of shifting to lower frequency (energy) upon pressure increase and higher frequency upon temperature increase, the modes shift to higher frequency upon pressure increase (shown later in Figure 5.8) and lower frequency upon temperature increase. This behaviour is the same as that of Raman frequencies in solids and can be understood using this analogy.

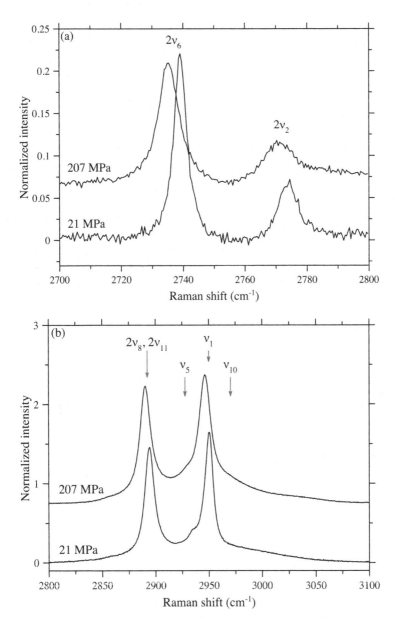

FIGURE 5.7 Raman spectra of liquid (subcritical) C_2H_6 at 300 K illustrating decrease in frequency of different vibrational modes upon pressure increase. Author's own data (further details given in ref. [3.24]).

We approach the topic using this method in Section 5.2.1. We review other approaches to modelling the peak position of vibrational Raman spectra in Section 5.2.2 and the peak position of rotational Raman spectra in Section 5.2.3. The intensity and linewidth of Raman spectra are discussed in Section 5.2.4. None of the approaches taken are perfect, and frequently experimental studies (e.g., ref. [1.9]) resort to purely empirical methods to model Raman shifts (frequencies) over a wide P,T range.

5.2.1 Grüneisen Model for Vibrational Raman Peak Position

We can begin our description of the Grüneisen model with the simplest case, an extended isotropic solid in which compression causes all interatomic distances to reduce in equal proportion. The frequency of Raman mode i is linked to the volume, and hence pressure, by the equation below [5.5]. The Raman frequency ω_i is linked to the volume V by the dimensionless constant γ_i, the Grüneisen parameter for this mode. The value of γ_i is typically small and positive, ca. 0.1–5.0. The value of γ_i is typically different for different Raman modes in the same material. The parameter ω_i^0 is the frequency of the mode at zero pressure (we earlier considered the effect of temperature in changing the volume):

$$\omega_i = \omega_i^0 \left[\frac{V(P)}{V_0} \right]^{-\gamma_i}$$

In this expression V_0 is the finite volume of the sample at zero pressure; the model does not allow for a phase transition to the gas state with $V \to \infty$ as $P \to 0$. We would like to model the Raman frequencies of dense molecular fluids and some modification to the equation is necessary to achieve this. Specifically, the assumption that all interatomic distances reduce in equal proportion upon compression is no longer reasonable, so we need to relate ω_i to the length $a(P)$ of the bond which is vibrating when mode i is excited (equation 5.14). This is a similar approach as that taken to model the Raman frequencies of highly anisotropic solids [5.10][5.11].

$$\omega_i = \omega_i^0 \left[\frac{a(P)}{a_0} \right]^{-\gamma_i'} \tag{5.14}$$

Here, a_0 is the bond length at zero pressure at the fixed non-zero temperature at which we are considering the Raman frequency and $a(P)$ is the bond length under pressure[s]. To complete our treatment we simply need to understand the origin of the Grüneisen parameter γ_i'. In the harmonic approximation, there would be no Raman frequency shift as a function of density, pressure, or temperature ($\gamma_i' = 0$), so the Grüneisen parameter measures the significance of anharmonic effects. These effects result from the deviation of the pair potential $\mathcal{V}(r)$ away from that of a simple harmonic oscillator, i.e., the potential not being completely symmetric about the equilibrium bond length. The

---

[s] Therefore $a_0 \neq r_0$ from equation 2.7.

methodology (resulting in equation 5.15) to derive γ_i' from the pair potential is given in various solid state physics textbooks (for example ref. [5.12]).

$$\gamma_i' = -\frac{r_0}{6} \times \frac{\frac{d^3\mathcal{V}}{dr^3}}{\frac{d^2\mathcal{V}}{dr^2}} \tag{5.15}$$

Here, r_0 is the equilibrium bond length at 0 K, which is also obtained from $\mathcal{V}(r)$. Hence the pair potential $\mathcal{V}(r)$ for the bond in question is the only information required to obtain the Grüneisen parameter. The most commonly used form for $\mathcal{V}(r)$ is the Lennard-Jones potential (equation 2.4 and Figure 2.1), in which case r_0 is as given in equation 2.7. This is correct for van der Waals bonding and works well for covalent bonding. In the cases of polar covalent and ionic bonding, the attractive part of the potential has a component varying $\propto 1/r$ due to the electrostatic attraction. The long range of this compared to the $\propto 1/r^6$ attractive part of the pure Lennard-Jones potential makes the potential more anharmonic, i.e., the potential is not so close to being symmetric about the equilibrium bond length. For the most general form of the pure Lennard-Jones potential (equation 5.16) we obtain $\gamma_i' = 3.5$ exactly, regardless of the value of the constants u_1, u_2. We would thus expect larger γ_i' for polar covalent or ionically bonded molecules.

$$\mathcal{V}(r) = \left(\frac{u_1}{r}\right)^{12} - \left(\frac{u_2}{r}\right)^{6} \tag{5.16}$$

Figure 5.8 shows the peak position of the principal Raman-active vibron of supercritical N_2 upon pressure increase at 300 K, commencing in the gas state then reaching liquid-like density upon pressure increase. The N–N bond length calculated from the peak position using equation 5.14 assuming $\gamma = 3.5$ is also given. As shown, the bond length can be measured with extremely high precision using this method, but the measurement could be subject to systematic errors.

5.2.2 Hard Sphere Fluid Theory of Vibrational Raman Peak Positions

In the preceding section, we obtained a relation between the vibrational Raman frequency and the relevant bond length but are left with the problem of how to model the complex variation in bond length as a function of pressure and temperature, beyond the simple observations that it initially increases as a function of density before decreasing at higher density. In the hard sphere fluid theory (as outlined and utilized in refs. [3.13] [5.13][5.14][5.15]) the total shift in Raman frequency $\Delta\omega$ as a function of pressure and temperature away from a certain reference P,T is instead divided into three separate components:

$$\Delta\omega = \Delta\omega_{att} + \Delta\omega_{rep} + \Delta\omega_{cen}$$

The term resulting from attractive forces between the molecules in the fluid, $\Delta\omega_{att}$, is negative and can be obtained theoretically subject to certain assumptions [3.13], leading

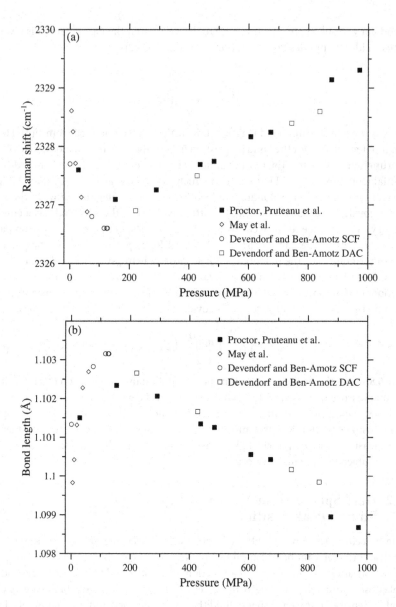

FIGURE 5.8 (a) Frequency of Raman-active vibration of fluid N_2 at 300 K, from ref. [4.8] and refs. therein. (b) Intramolecular N–N bond length calculated from the Raman frequency using equation 5.14.

to a shift that varies linearly with density and thus dominates at low density compared to higher-order terms (varying as ρ^2, ρ^3, etc.). The term $\Delta\omega_{rep}$ resulting from repulsive forces is obtained from a pair potential $\mathcal{V}(r)$ with a particle diameter obtained from a P,V,T EOS of the Carnahan-Starling-van der Waals form. The hard sphere diameter of the fluid molecules is an input parameter which is varied to obtain an EOS fitting the data over a wide P,V,T range. The last term, the centrifugal term $\Delta\omega_{cen}$, is a small negative offset. It is linearly temperature dependent and reflects the slight loosening of the intramolecular bonds resulting from the fluid molecules existing in rotational excited states upon temperature increase.

5.2.3 Peak Position of Rotational Raman Spectra

The rotational Raman modes of most molecules lie at frequency (energy) too low to be detectable with a conventional (single-grating) Raman spectrometer as they are obscured by the notch filters used to cut out the Rayleigh-scattered and reflected light. They are observable using triple-grating spectrometers but few of these instruments exist, and the observed scattering intensity is often very weak. Thus there are relatively few studies of pure rotational Raman modes of fluids covering a reasonable P,T range.

An important exception to this is H_2 as it has rotational Raman modes lying at unusually high energy, ca. 350–600 cm^{-1} (435–746 meV) detectable on a single-grating Raman spectrometer. The leading-order term determining the rotational energy levels is [5.16][5.17]:

$$E_{rot}(J) = \frac{\hbar^2}{2\mu} \times \frac{1}{r^2} \times J(J+1)$$

Here, μ is the reduced mass of the H_2 molecule, r is the separation between the two atomic centres in the molecule, and J is the quantum number for the relevant excited state. The next term accounts for centrifugal effects and is nearly an order of magnitude smaller. It is thus possible to estimate r with reasonable accuracy from the rotational transition frequencies [5.16].

5.2.4 Peak Intensity and Linewidth of Fluid Raman Spectra

Measuring the absolute intensity of Raman peaks is generally not possible but the changes in intensity as pressure and temperature are varied can be measured with reasonable accuracy as long as care is taken. Two approaches have been taken to modelling the variation in the integrated peak intensity I as a function of density. The first approach [5.18] is a semi-empirical one in which the intensity is expanded as a power series in terms of the density:

$$I = A\rho + B\rho^2 + C\rho^3$$

In this case the first term represents what we intuitively expect from increasing the density: The more molecules are present in the laser beam (which has a fixed diameter at the focal point) the greater the intensity, assuming there is no effect of interaction between the molecules. The second term represents increases in intensity taking place through interaction of two molecules, then the third through interaction of three molecules.

The second approach taken relates the intensity directly to the dynamic susceptibility $\chi(\omega)$. Since the absolute intensity is not known, the best we can do [5.19][5.20][5.21] is write (for the intensity of the Stokes peaks):

$$I(\omega) \propto (\omega_L - \omega)^4 \left(\frac{1}{e^{\frac{\hbar\omega}{k_B T}}} - 1 \right) \chi(\omega)$$

Here, ω_L is the frequency of the incident laser radiation and ω is the frequency of the vibration/rotation being excited (*not* the absolute frequency of the scattered photon). Unfortunately, the variation of $\chi(\omega)$ as a function of temperature and density is not generally known from first principles so this method is not widely used. An additional Bose-Einstein factor appears if we wish to calculate the intensity of the anti-Stokes peaks.

The linewidths observed in fluid Raman spectra are potentially of great importance for future study since there is evidence from a few recent studies that the linewidth exhibits fundamentally different behaviour for gas-like and liquid-like fluids [3.24][4.8] [4.17], particularly in the supercritical region where the gas-like and liquid-like regions are demarcated by the Frenkel line (Section 6.5).

In the gas state, linewidths initially increase upon density increase due to collision broadening. This effect is observed at extremely low densities that are not of interest to us in this text. Then, at higher density we observe collisional narrowing instead ([5.22] and refs. therein). However, at even higher density the peaks once again broaden upon density increase [3.24][4.8][4.17]. The exact mechanism for this is still open for debate, but one strong candidate is the ability of dense fluids as discussed earlier to exhibit solid-like behaviour on timescales shorter than the liquid relaxation time. Solids can support some amount of static nonhydrostatic stress. As a result, the pressure in a solid can vary spatially and (at a given location) can be different along different axes. Both these phenomena cause Raman peaks to broaden.

A dense fluid, that we will refer to from Chapter 6 onwards as being a rigid liquid, cannot support a static nonhydrostatic stress but can support nonhydrostatic stress on timescales shorter than the liquid relaxation time. Thus the broadening of spectral peaks upon density increase can occur. Figure 5.9 shows the trend in peak width upon pressure increase at 300 K for N_2 and CH_4, illustrating in both cases the change between the two different regimes—collisional narrowing at lower density and broadening due to nonhydrostatic stress at higher density.

Typically, widths of Raman peaks from fluids are significantly lower than from solids. The measured peak widths from fluids may therefore not be quantitatively accurate or comparable between different studies since they are convoluted with, and usually of similar magnitude to, the intrinsic linewidth set by the spectral resolution of the spectrometer.

5.2.5 Prediction of Fluid Raman Spectra Using MD

Above we have outlined various simple analytical, qualitative, or semi-empirical approaches to predicting fluid Raman spectra. Alternatively, one could attempt to

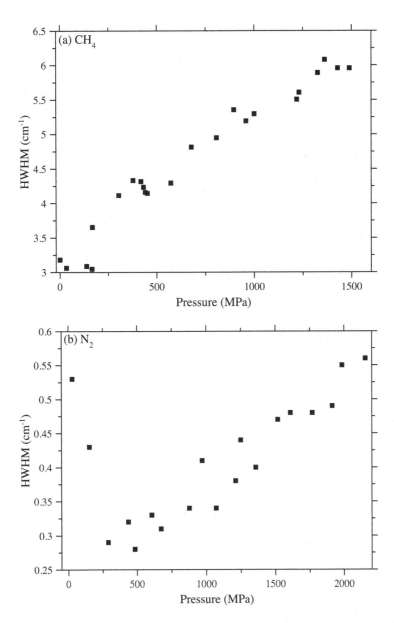

FIGURE 5.9   Half width half maximum (HWHM) of vibrational Raman peaks from (a) CH$_4$ and (b) N$_2$ collected upon pressure increase at 300 K. Authors own data (adapted from refs. [4.17] and [4.8]).

predict the form of the Raman spectra using ab initio MD. This approach is routine for solids but is in its infancy for fluids. Presently it is possible to predict the Raman peak positions with reasonable accuracy (for instance ref. [4.8]). However, this is only adequate for the very simplest fluids, as in most cases the Raman spectra consist of groups of overlapping peaks. Thus, to model these spectra using MD it is necessary to predict with reasonable accuracy the linewidth and relative intensity of the peaks in addition to their positions.

5.3 Brillouin Spectroscopy of Liquids Close to the Melting Curve

We have seen that the ability to support shear waves is an archetypal solid-like property. The ability of dense liquids to do this also has been directly verified by various research groups over the years using Brillouin spectroscopy [5.23][5.24][5.25][5.26][5.27] (for example the spectra shown in Figure 5.10 (a)) as well as with other methods as noted in Section 5.1.2. Brillouin spectroscopy measures the inelastic scattering of photons (usually produced by a visible laser) by acoustic waves (in a solid, and in a dense fluid, we can refer to these excitations as phonons). We begin by outlining how this process works in terms of the diffractive effect of long-wavelength phonons. We can consider (Figure 5.10 (b)) the part of the phonon dispersion relation covering longitudinal and

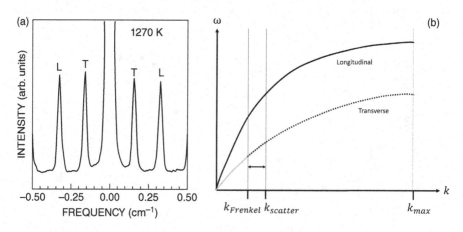

FIGURE 5.10 (a) Brillouin spectrum from liquid B_2O_3 at 1270 K showing the existence of transverse/shear (T) as well as longitudinal (L) modes [5.27]. Panel (a) is reprinted with permission from M. Grimsditch, R. Bhadra and L.M. Torell, Phys. Rev. Lett. **62**, 2616 (1989). Copyright (1989) by the American Physical Society. (b) Schematic diagram of a fluid phonon dispersion relation showing longitudinal and transverse modes. In the case of the transverse modes, the grey solid line indicates the modes that are detectable using Brillouin scattering and the black dotted line indicates the modes that propagate in the fluid. The wavenumber $k_{Frenkel}$ marks the lowest wavenumber transverse modes that exist (also at the lowest frequency ω) whilst $k_{scatter}$ marks the highest wavenumber mode that can be excited by an incident photon (consistent with the fundamental selection rule).

transverse acoustic phonons in the low-energy limit. In the longitudinal *and* transverse cases, as the $k \rightarrow 0$ limit is approached the wavelength increases until it becomes comparable to the wavelength of the incident laser beam. At this point the sample essentially acts as a diffraction grating leading to scattering of the incident light. Acoustic phonons can be created (Stokes) or annihilated (anti-Stokes) in the scattering process.

There are some similarities and differences between this process and the Raman scattering process described earlier. The key similarity is that both these *inelastic* scattering processes have a vanishingly small intensity compared to that of the elastic Rayleigh scattering process. We can write down, for both Raman and Brillouin spectroscopy, the consequences of the conservation of energy and momentum in the process (equation 5.17). The phonon frequency is ω_{ph} and the wavevector is k_{ph}, similarly the parameters for the laser and scattered photons are defined.

$$\hbar \omega_s = \hbar \omega_l \pm \hbar \omega_{ph} \tag{5.17}$$

$$k_s = k_l \pm k_{ph}$$

The first difference between Raman and Brillouin spectroscopy is that the phonon energy in the Brillouin scattering process is tiny compared to that in the Raman scattering process. Hence thermal energies are always sufficient to excite the anti-Stokes peak in Brillouin scattering but are usually not sufficient in Raman scattering. In addition, in Brillouin scattering, due to the spectral proximity of the Brillouin-scattered photons to the laser wavelength, different (read: more expensive) apparatus is needed to filter out the elastically scattered light and measure the wavelength of the Brillouin-scattered photons. A Fabry-Perot interferometer is required, instead of a spectrometer operating via dispersion at a diffraction grating(s). The shear modes in the fluid have the lowest energy (Figure 5.10) and are hence the hardest to detect. In fact, the form of the dispersion relation shown in Figure 5.10 (b) in which there exist, in the low-frequency/ hydrodynamic limit, separate dispersion curves for longitudinal (higher energy) and transverse (lower energy) modes is a fairly general result. In addition to its derivation for fluids [1.11] it is also obtained, for instance, for propagation of elastic waves in a generic isotropic medium [5.28] or a monatomic three-dimensional lattice [2.7].

Secondly, the mechanism for scattering to occur is completely different. The Raman scattering is due to the change in polarizability when an optical phonon is excited. The Brillouin scattering, on the other hand, is due to the scattering mechanism of the excited sample acting as a diffraction grating. The scattering therefore takes place at certain specific angles. Optical access to the sample from a variety of directions and a transparent sample is required. The information that can be obtained from an opaque sample in the backscattering geometry is limited. In particular, observation of the transverse/shear modes is not possible in the backscattering geometry. In Raman scattering, in contrast, we usually utilize the backscattering geometry for simplicity.

The dependence of signal on scattering angle leads to Brillouin scattering providing additional information about the excited phonons. Since the exact vectorial values of both the incident and scattered light wavevectors are known, we can use equation 5.17 to calculate the wavevector of the phonon as well as the energy. In this case the phonon group velocity v can also be calculated ($v = \omega/k$) and elastic moduli can be calculated if the P, V EOS is known [5.29].

The scientific literature does not contain a large number of Brillouin spectroscopy studies mapping out shear wave propagation in dense fluids in a systematic way. Partly, this is because Brillouin spectroscopy cannot detect the full spectrum of shear waves expected in dense fluids. Only the lowest frequency shear waves in the most viscous fluids can be detected. The full spectrum of shear waves consists of waves with wavevector k ranging from zero up to the boundary of the first pseudo-Brillouin zone. The maximum value of k is therefore of the order of magnitude $k_{max} \sim 10^{10} \mathrm{m}^{-1}$. However, for Brillouin scattering to occur the wavevector of the wave needs to be similar to that of the incident laser light, or smaller, dictating that we can detect waves only up to $k_{scatter} \sim 10^7 \mathrm{m}^{-1}$ using this method.

For shear waves to exist with such (relatively speaking) small wavenumber, the Frenkel frequency ω_F and wavevector k_F need to be reduced so they are virtually at the centre of the pseudo-Brillouin zone. Referring to equation 5.12, we can see that to achieve this the viscosity needs to be as high as possible. Thus successful Brillouin spectroscopy experiments cited earlier have focussed on the most viscous fluids. Figure 5.10 (b) illustrates schematically an example of a dispersion relation where this condition is satisfied. The arrow indicates the range of shear modes that propagate and can be detected.

5.4 Inelastic Neutron and X-ray Scattering from Liquids Close to the Melting Curve

5.4.1 Distinction Between Neutron and X-ray Scattering

Just as with coherent elastic scattering, there are certain key differences between X-ray and neutron coherent inelastic scattering. In both cases we seek to use a probe with a wavelength similar to the interatomic spacing in the fluid, resulting in very different probe energy in the X-ray and neutron case. With $\lambda = 1$ Å we obtain:

$$E = hc / \lambda = 12.4 \text{ keV (X-rays)} \tag{5.18}$$

$$E = h^2 / 2m_n \lambda^2 = 0.08 \text{ eV (neutrons)} \tag{5.19}$$

The energy of excitations such as sound wave quanta that we study using inelastic X-ray and neutron scattering is ~0.01 eV, so the change in energy of a neutron on creation or annihilation of such an excitation is a significant proportion of the total neutron energy. For this reason, study of fluids using inelastic neutron scattering has been performed for several decades whilst similar studies using inelastic X-ray scattering have only become possible recently.

However, the much higher energy of the incident probe compared to the excitations under study in the case of inelastic X-ray scattering allows a far larger area of the dispersion relation to be measured. Writing down the conservation of momentum and energy in an inelastic X-ray scattering event in which the incident X-ray or neutron with initial properties E, ω, k loses energy $\hbar\Delta\omega$ and undergoes a change in wavevector Q we obtain (using the notation of Figure 4.4 except that $k \neq k'$ now):

$$Q = k' - k \qquad (5.20)$$

$$E = \hbar\omega = \hbar\Delta\omega + \hbar\omega' \qquad (5.21)$$

For both neutrons and X-rays we may square equation 5.20 to obtain:

$$\left(\frac{Q}{k}\right)^2 = 1 + \left(\frac{k'}{k}\right)^2 - 2\left(\frac{k'}{k}\right)\cos 2\theta \qquad (5.22)$$

Here, 2θ is the scattering angle obtained from $k.k'$. Obtaining the ratio of the energy loss to the initial energy using equation 5.21 gives us, using the X-ray dispersion relation $\omega = ck$:

$$\frac{\hbar\Delta\omega}{E} = 1 - \frac{k'}{k}$$

Since, for inelastic X-ray scattering $E \gg \hbar\Delta\omega$, we obtain $k' \approx k$ and equation 5.22 simplifies to:

$$\left(\frac{Q}{k}\right)^2 = 1 - 2\cos 2\theta$$

Thus, by varying θ we can access a large range of Q values regardless of the value of $\Delta\omega$, within the constraint $E \gg \hbar\Delta\omega$. The conservation rules for energy and momentum expressed in equations 5.20 and 5.21 do not put any significant constraint on our ability to map out dispersion relations using inelastic X-ray scattering.

On the other hand, dividing equation 5.21 by E and applying the free neutron dispersion relation $\omega = \hbar k^2 / 2m_n$ instead we obtain:

$$\frac{\hbar\Delta\omega}{E} = 1 - \left(\frac{k'}{k}\right)^2$$

Here, we may no longer make the approximation $E \gg \hbar\Delta\omega$ and substitution into equation 5.22 yields:

$$\left(\frac{Q}{k}\right)^2 = 2 - \frac{\hbar\Delta\omega}{E} - 2\left(1 - \frac{\hbar\Delta\omega}{E}\right)^{\frac{1}{2}}\cos 2\theta$$

This is a quadratic function of $\left(1 - \frac{\hbar\Delta\omega}{E}\right)^{\frac{1}{2}}$, so it can be solved using the common formula to obtain:

$$\hbar\Delta\omega = E\left[1 - \cos 2\theta \pm \sqrt{\left(\frac{Q}{k}\right)^2 - \left(\sin 2\theta\right)^2}\right] \qquad (5.23)$$

Thus, in this case the energy and momentum conservation rules place strict constraints on what values of $\Delta\omega$ can be accessed for a given value of Q. In the most extreme case of $Q=0$ we can only access excitations with $\hbar\Delta\omega=0$, at $2\theta=0$. As Q increases a range of different values of $\hbar\Delta\omega$ become accessible by varying 2θ (Figure 5.11 (a)) allowing the shaded part of the dispersion relation in Figure 5.11 (b) to be accessed.

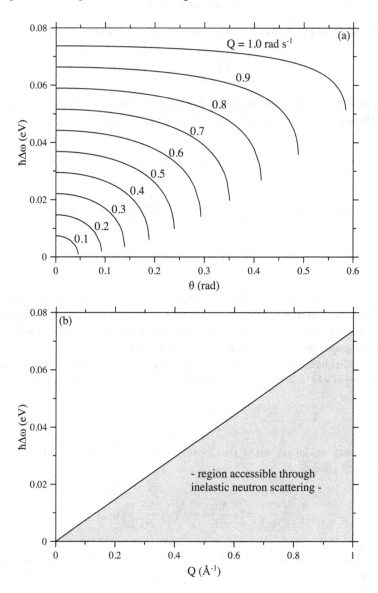

FIGURE 5.11 (a) Values of $\hbar\Delta\omega$ as a function of neutron scattering angle for various scattering vectors Q. (b) Maximum $\hbar\Delta\omega$ accessible as a function of Q, in the $2\theta\to0$ limit. Data were calculated using $k=1.0854$ Å^{-1} for the incident neutrons, resulting in $E=0.08$ eV.

5.4.2 The Scattered Intensity

Whilst elastic neutron and X-ray scattering (diffraction) from fluids provides information about the static structure only, inelastic scattering provides information about the dynamic properties. In particular, it provides the means to measure the dispersion relation for a far greater proportion of the phonons in the fluid, without the limitations of Brillouin and Raman scattering.

A full mathematical treatment of coherent inelastic neutron and X-ray scattering from fluids, leading to the dynamic structure factor, is lengthy and the present author has nothing new to say about it. Since it is given in many other texts and electronic resources, for instance refs. [1.11][4.11][5.30], we shall simply quote the result here, using notation consistent with that employed elsewhere in this text:

$$I_{coh}(\mathbf{Q},\omega)=\frac{1}{h}\frac{k'}{k}\iint\left[\sum_{m=1,n=1}^{m=N,n=N}b_m b_n \delta\big[\mathbf{r}-\big(\mathbf{r}_n(0)-\mathbf{r}_m(t)\big)\big]\right]e^{-i\mathbf{Q}\cdot\mathbf{r}}e^{-i\omega t}d^3 r dt \quad (5.24)$$

Here, $I_{coh}(\mathbf{Q},\omega)$ is the scattered intensity as a function of the scattering vector $\mathbf{Q}$ and the frequency ω corresponding to the change in energy of the neutron/X-ray upon scattering into the selected energy range**. The wavenumbers k, k' are the scalar values of the wavevectors $\mathbf{k}, \mathbf{k}'$ defined in Figure 4.4 but are no longer equal in magnitude as the scattering may be inelastic. The equation as shown is for neutron scattering with the scattering lengths b_i of the different atoms included. For X-ray scattering we would substitute the Q-dependent atomic form factors $f_i(Q)$. The parameter $\mathbf{r}_n(0)$ gives the position of the nth atom at $t=0$ and $\mathbf{r}_m(t)$ gives the position of the mth atom at t. The double summation is over all m,n including the terms where $m=n$.

Elsewhere (Chapter 4 and Appendix C) we have understood elastic X-ray and neutron diffraction from fluids as providing us with the Fourier transform of the function $g(r)$ representing the spatial distribution of the atoms. For the study of dynamic properties we cannot consider simply a snapshot of the particles at a single point in time, we need to study the evolution of their positions as a function of time. Thus the double integral in equation 5.24 represents a double Fourier transform over all space and all time. The time taken to obtain an X-ray or neutron diffraction pattern is certainly extremely large compared to the timescales for excitations on a microscopic scale.

The terms within the integral in equation 5.24 are represented as a parameter (normalized with respect to N) called the dynamic structure factor $S(\mathbf{Q},\omega)$, in contrast to the static structure factor which is dependent on $\mathbf{Q}$ only. For a fluid composed of identical atoms $b_i \to b$ and we obtain:

$$I_{coh}(\mathbf{Q},\omega)=\frac{Nb^2}{h}\frac{k'}{k}S(\mathbf{Q},\omega)$$

** In an inelastic X-ray / neutron scattering experiment, the scattered X-rays / neutrons are collected at a specified scattering angle / vector θ / $\mathbf{Q}$ (just as in an elastic scattering experiment) but are then dispersed according to their energy distribution at the selected scattering angle.

Where:

$$S(\mathbf{Q}, \omega) = \frac{1}{N} \iint \left[\sum_{m=1, n=1}^{m=N, n=N} \delta \left[r - \left(r_n(0) - r_m(t) \right) \right] \right] e^{-i\mathbf{Q} \cdot r} e^{-i\omega t} d^3 r \, dt \qquad (5.25)$$

This expression includes the (dominant in terms of intensity) contribution from elastic scattering and is consistent with our definition of $S(Q)$ in equations 4.3–4.5. If we assume that all atoms remain in their equilibrium positions then the only time dependence remaining in the expression for $S(\mathbf{Q}, \omega)$ is the $e^{-i\omega t}$ term which integrates with respect to time to give a δ-function centered on $t = 0$. We therefore obtain:

$$S(\mathbf{Q}) = \frac{1}{N} \int \left[\sum_{m=1, n=1}^{m=N, n=N} \delta \left[r - (r_n - r_m) \right] \right] e^{-i\mathbf{Q} \cdot r} d^3 r$$

Separating the summation out into the terms for which $m = n$ and $m \neq n$ we obtain:

$$S(\mathbf{Q}) = 1 + \frac{1}{N} \int \left[\sum_{\substack{m=1, n=1 \\ m \neq n}}^{\substack{m=N, n=N}} \delta \left[r - (r_n - r_m) \right] \right] e^{-i\mathbf{Q} \cdot r} d^3 r$$

The remaining terms in the summation now consist of δ-functions centered on all values of r that correspond to a separation between two particles in the system. Each term is then multiplied by the phase factor $e^{-i\mathbf{Q} \cdot r}$ to account for the interference between the coherently scattered beams from each particle at the chosen value of $\mathbf{Q}$. For a spherically symmetric system this can be averaged over different orientations to obtain the expression utilized for elastic scattering in equations 4.3–4.6.

The coherent inelastic scattering in which we are usually interested takes place due to scattering events where an excitation such as a low-frequency sound wave is excited/annihilated (in a solid we would refer to these as phonons). These appear as weak sidebands on the elastic scattering peaks at selected Q values as shown, for instance, in Figure 5.12. This figure shows inelastic X-ray scattering (IXS) spectra of fluid O_2 at 300 K and various pressures; a representative selection of the large number of spectra required to map out the dispersion curves for (in this case) the low energy acoustic phonons. Each individual spectrum corresponds to the different energy X-rays scattered at a selected value of Q. Thus the dispersion relation can be mapped out from a large number of spectra collected at the same pressure and different values of Q.

Data collection times for inelastic neutron and X-ray scattering data are therefore extremely long. It is challenging to maintain fluids at extreme but constant P, T conditions for these times and challenging to secure enough beamtime at the small number

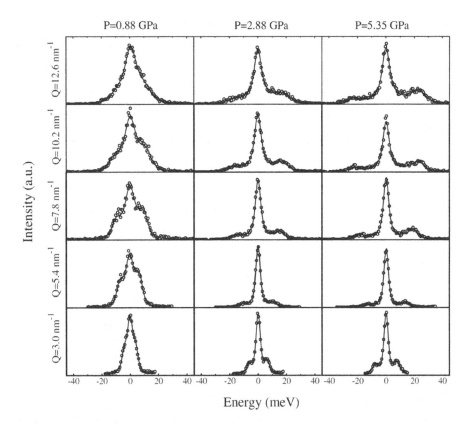

FIGURE 5.12   Selected IXS spectra of fluid oxygen at elevated pressures and 300 K [5.31]. Reprinted figure with permission from F. Gorelli, M. Santoro, T. Scopigno, M. Krisch and G. Ruocco, Phys. Rev. Lett. **97**, 245702 (2006). Copyright (2006) by the American Physical Society.

of neutron sources and synchrotron X-ray sources worldwide where such experiments can be performed. These factors make it hard to map out phonon dispersion relations in fluids across wide areas of P, T space.

5.4.3 What Can We Learn from Inelastic Neutron and X-ray Scattering from Liquids?

We can thus map out dispersion curves for phonon-like excitations in liquids (and dense supercritical fluids) in an analogous way to the dispersion curves for solids. The key phenomenon in liquids which can be studied using inelastic neutron and X-ray scattering from liquids is *positive sound dispersion* (PSD). The PSD phenomenon is the existence of two fundamentally different regimes in the dispersion relation for low-frequency, low-wavevector acoustic excitations, depending on the frequency ω. For the very lowest frequencies, referred to as the viscous regime, the liquid is able to respond completely to the perturbation of each wavefront arriving. If we denote the time taken by the system to respond as a structural relaxation time τ_R then we can define the viscous regime using the condition:

$$\omega\tau_R \ll 1$$

At the other extreme, when ω becomes sufficiently high that the system does not have time to respond to each incoming wavefront, we are in the elastic regime in which:

$$\omega\tau_R \gg 1$$

It is valid to describe the liquid's behaviour in the elastic regime as solid-like, since the liquid does not have time to exhibit its ability to flow between the arrival of successive wavefronts. Figure 5.13 illustrates schematically the response of the liquid to a hypothetical square wave in both extreme cases.

If we measure the dispersion relation for the acoustic waves in a dense liquid, we observe a fundamentally different dispersion relation in the viscous and elastic regimes with a crossover between these regimes at some intermediate frequency. This observation is known as positive sound dispersion (PSD) and is an important characteristic property of liquid, and liquid-like, states of matter. Figure 5.14 is an example of an observation of PSD, by means of inelastic X-ray scattering from liquid Ne. We shall discuss in Section 6.5.3 the observation of PSD in supercritical fluids at liquid-like density.

Completely separate from the viscous to elastic transition, and lying generally at higher frequency, is the adiabatic to isothermal transition. Upon propagation of a longitudinal sound wave we expect an increase in temperature in the compressed parts of the medium through which the wave propagates and decrease in temperature in the rarefied parts. The wave propagation can be adiabatic (if there is not enough time for heat to be transferred to equalize temperature before the next wavefront arrives) or isothermal (if the system is able to reach thermal equilibrium between the arrival of successive wavefronts). A thermal relaxation time can be defined to demarcate these two regimes; different to the structural relaxation time discussed earlier.

Generally, propagation of sound waves is adiabatic [5.33]. As we increase the frequency and wavevector two competing effects influence the balance between adiabatic and isothermal wave propagation: The increase in frequency makes it harder for the system to reach thermal equilibrium between the arrival of successive wavefronts; on the other hand, the decrease in wavelength makes it easier. The latter effect dominates and at high frequency a transition from adiabatic to isothermal wave propagation occurs. Generally this is at a much higher frequency than the viscous–elastic transition but this is not necessarily the case; the two transitions can overlap [5.34]. Importantly, observation of the adiabatic–isothermal transition is not proof that the sample is in a liquid or liquid-like state. The adiabatic to isothermal transition is also observed in gases [5.35][5.36].

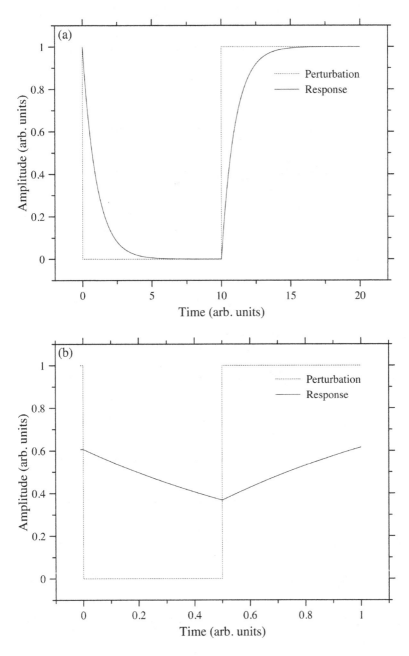

FIGURE 5.13   Schematic of sample response to perturbing square sound wave in the viscous (a) and elastic (b) regimes.

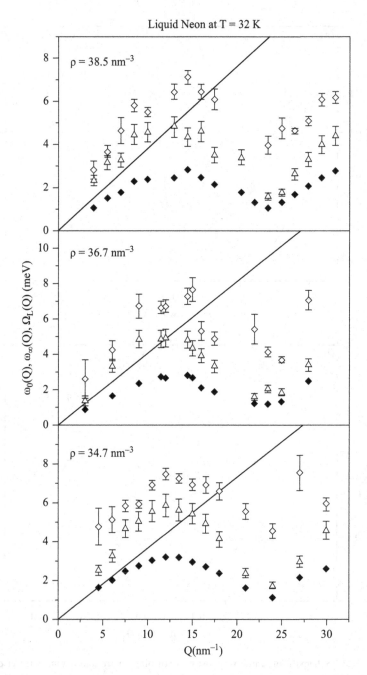

FIGURE 5.14 Dispersion relation for acoustic excitations in liquid Ne at different densities [5.32], demonstrating the presence of PSD. Reprinted from A. Cunsolo, G. Pratesi, R. Verbeni, D. Colognesi, C. Masciovecchio, G. Monaco, G. Ruocco and F. Sette, J. Chem. Phys. **114**, 2259 (2001), with the permission of AIP Publishing.

5.5 Summary and Outlook

The study of the dynamic properties of dense liquids and supercritical fluids reflects their intermediate position between gases and solids, similarly to their static properties. This is explicitly accounted for in the Frenkel model, in which the ability of the fluid to exhibit rigidity (elastic response to applied stress) and viscosity on different timescales is encapsulated in an atomistic manner in the parameter of the liquid relaxation time. Although dating back many decades [5.1], it is only since 2012 that the Frenkel model has been used to predict the heat capacities of fluids. This is the first time that a reasonably accurate method to predict fluid heat capacity has been based reasonably closely on first principles.

Most recently, it has been proposed that the propagation (or lack thereof) of the shear waves in fluids is dictated by their wavenumber rather than frequency. Instead of the lowest frequency modes being unable to propagate it is the lowest wavenumber modes [5.37][5.38] (the so-called k-gap). It is agreed that the dispersion for transverse waves can be described using the following relation (given in refs. [1.11][5.37] [5.38][5.39]):

$$\omega(k) = \sqrt{c_T^2 k^2 - \frac{1}{\tau_R^2}}$$

(5.26)

Here, c_T is the speed (group velocity) of the wave and τ_R is a relaxation time. Below a certain cutoff wavevector (the k-gap) the frequency is imaginary. However, the extent to which the Maxwell and Frenkel relaxation times can be regarded as similar and equated to τ_R remains controversial [5.39]. In the simplest approximation (that the expression for the energy from equation 5.7 can be substituted with $U_{ph} = k_B T$) the total internal energy can be obtained using the density of states as a function of k and equation 5.26 is used to determine the range of integration. This gives the same final expression for the internal energy as the analogous procedure with the frequency gap [5.37]. However, a fully quantum-mechanical version of this derivation, applied to model the internal energy and heat capacities of real fluids, has yet to be investigated to the author's knowledge.

The heat capacities of key fluids over a wide P, T range, while poorly understood, are well characterized experimentally. However, the study of the excitations in fluids on a microscopic scale is more complex. Excitations within individual molecules in molecular fluids, and collective phonon-like excitations in all fluids, are studied principally utilizing Raman spectroscopy (individual molecule vibrations/rotations) and inelastic neutron/X-ray scattering (collective excitations such as sound waves). Both these excitations behave in a fundamentally different manner in dense fluids compared to gases and even to liquids near to the critical point.

In the next chapter we apply this understanding to elucidating the behaviour of supercritical fluids over the large ρ, T range covered by the supercritical region for different fluids.

References

5.1 Y.I. Frenkel, *Kinetic theory of liquids*, Dover publications, New York, (1955).

5.2 T. Pezeril, C. Klieber, S. Andrieu and K.A. Nelson, Phys. Rev. Lett. **102**, 107402 (2009).

5.3 S. Hosokawa et al., Phys. Rev. Lett. **102**, 105502 (2009).

5.4 Y.H. Jeong, S.R. Nagel and S. Bhattacharya, Phys. Rev. A **34**, 602 (1986).

5.5 B. Weinstein, and R. Zallen, Pressure-raman effects in covalent and molecular solids, in *Light scattering in solids IV*, M. Cardona, and G. Güntherodt, (eds.), Springer, New York (1984).

5.6 C.M. Carlson, H. Eyring and T. Ree, PNAS **46**, 649 (1960).

5.7 C. Grimvall, Physica Scripta **11**, 381 (1975).

5.8 H. Nakayama, K.-I. Saitow, M. Sakashita, K. Ishii and K. Nishikawa, Chem. Phys. Lett. **320**, 323 (2000).

5.9 V.G. Arakcheev, V.N. Bagratashvili, A.A. Valeev, V.B. Morozov and V.K. Popov, Rus. J. Phys. Chem. B **4**, 1245 (2010).

5.10 M. Hanfland, H. Beister and K. Syassen, Phys. Rev. B **39**, 12598 (1989).

5.11 J.E. Proctor, E. Gregoryanz, K.S. Novoselov, M. Lotya, J.N. Coleman and M.P. Halsall, Phys. Rev. *B* **80**, 073408 (2009).

5.12 J.R. Hook and H.E. Hall, *Solid state physics*, Wiley, New York (1991).

5.13 M.-R. Lee and D. Ben-Amotz, J. Chem. Phys. **99**, 10074 (1993).

5.14 J. Zhang, S. Qiao, W. Lu, Q. Hu, S. Chen and Y. Liu, J. Geochem. Exp. **171**, 20 (2016).

5.15 D. Ben-Amotz, M.-R. Lee, S.Y. Cho and D.J. List, J. Chem. Phys. **96**, 8781 (1992).

5.16 N.E. Moulton, G.H. Watson Jr., W.B. Daniels and D.M. Brown, Phys. Rev. A **37**, 2475 (1988).

5.17 J.M. Hollas, *Modern spectroscopy*, Wiley, New York (2004).

5.18 C.H. Wang and R.B. Wright, J. Chem. Phys. **59**, 1706 (1973).

5.19 H. Nakayama, S. Yajima, T. Yoshida and K. Ishii, J. Raman Spec. **28**, 15 (1997).

5.20 H. Nakayama, K.-I. Saitow, M. Sakashita, K. Ishii and K. Nishikawa, Chem. Phys. Lett. **320**, 323 (2000).

5.21 W. Hayes and R. Loudon, *Scattering of light by crystals*, Wiley-VCH, New York (1978).

5.22 A.D. May, J.C. Stryland and G. Varghese, Can. J. Phys. **48**, 2331 (1970).

5.23 G.I.A. Stegeman and B.P. Stoicheff, Phys. Rev. Lett. **21**, 202 (1968).

5.24 V. Volterra, Phys. Rev. **180**, 156 (1969).

5.25 G.D. Enright and B.P. Stoicheff, J. Chem. Phys. **64**, 3658 (1976).

5.26 C.H. Wang, Mol. Phys. **41**, 541 (1980).

5.27 M. Grimsditch, R. Bhadra and L.M. Torrell, Phys. Rev. Lett. **62**, 2616 (1989).

5.28 L.D. Landau and E.M. Lifshitz, *Theory of elasticity*, Pergamon Press, Oxford (1969).

5.29 A. Polian, J. Raman Spec. **34**, 633 (2003).

5.30 F.A. Gorelli, Inelastic x-ray scattering on high-pressure fluids, in *high-pressure physics*, J.S. Loveday (ed.), CRC Press, Boca Raton (2012).

5.31 F. Gorelli, M. Santoro, T. Scopigno and G. Ruocco, Phys. Rev. Lett. **97**, 245702 (2006).

5.32 A. Cunsolo et al., J. Chem. Phys. **114**, 2259 (2001).

5.33 L.E. Kinsler, A.R. Frey, A.B. Coppens and J.V. Sanders, *Fundamentals of acoustics*, Wiley, New York (2000).

5.34 *The high-frequency dynamics of liquids and supercritical fluids*. Ph.D. thesis of Filippo Bencivenga, Université Joseph-Fourier (2006). https://tel.archives-ouvertes.fr/tel-00121509/

5.35 J. Wu. Am. J. Phys. **58**, 694 (1990).

5.36 N.H. Fletcher, Am. J. Phys. **42**, 487 (1974).

5.37 C. Yang, M.T. Dove, V.V. Brazhkin and K. Trachenko, Phys. Rev. Lett. **118**, 215502 (2017).

5.38 M. Baggioli, M. Vasin, V.V. Brazhkin and K. Trachenko, Phys. Rep. **865**, 1 (2020).

5.39 T. Bryk, I. Mryglod, G. Ruocco and T. Scopigno, Phys. Rev. Lett. **120**, 219601 (2018).

6

Beyond the Critical Point

At the critical point, the first order phase transition between the liquid and gas states ends. There are no other first order phase transitions present at higher P,T until the melting curve is approached. However, there are a number of narrow transitions (that is, changes occurring in a narrow P,T range) in both static and dynamic properties extending deep into the supercritical region. Some examples (for instance the Widom lines) have been studied experimentally for decades, whilst the Frenkel line, on the other hand, has been proposed only recently. In this chapter we will outline the theoretical aspects of these transitions and review the experimental evidence for their existence. We will begin with the Widom lines, then study the Fisher-Widom line, the various inversion curves, and finally the Frenkel line.

6.1 The Widom Lines

The vapour pressure curve, marking the first order phase transition between the liquid and gas states, consists of discontinuous changes in various thermodynamic functions. It ends at the critical point. However, all functions that undergo the first order phase transition when the vapour pressure curve is crossed also undergo a narrow transition when a P,T path is followed just beyond the critical point, roughly following an extrapolation of the vapour pressure curve along an isochore. For instance, there is a discontinuous change in the density ρ and viscosity η when the vapour pressure curve is crossed, resulting in a spike in the compressibility and $d\eta / dP$. Along a P,T path just beyond the critical point crossing an extrapolation of the vapour pressure curve there is still a maximum (or minimum) in these parameters, becoming progressively more smeared out further away from the critical point. A Widom line is a line in P,T space linking these extremal values for a given thermodynamic function.

All condensing fluids exhibit Widom lines; for example, a variety of Widom lines can be plotted out beginning from the simplest P,V,T EOS for a condensing fluid, the van der Waals EOS. Figure 6.1 shows the Widom line for compressibility arising from the van der Waals EOS extending from $T = T_C$ to $T \approx 1.5T_C$. Other Widom lines for the van der Waals fluid are given in ref. [6.1]. Some texts describe the Widom lines as second order phase transitions. In the author's view, inspection of data such as Figures 6.1–6.4 demonstrates that this is not the case. The definition of a second order phase transition is that the derivative of the property in question exhibits a discontinuity.

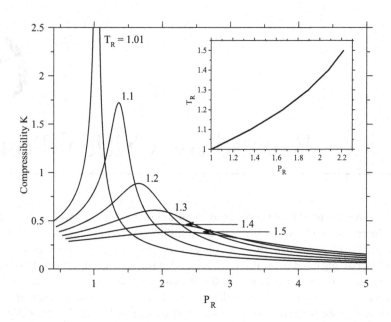

FIGURE 6.1   Compressibility maximum (Widom line) plotted for the van der Waals fluid from $T/T_C = T_R = 1.01-1.5$. At $T_R = 1.5$ the maximum has virtually disappeared. Inset: Plot of the compressibility maximum points on the P,T phase diagram.

Whilst the derivative of (for example) the compressibility does change sign upon pressure (or temperature) change crossing the Widom line, there is no discontinuity and therefore no second order phase transition.

A Widom line can be plotted for any function that undergoes a discontinuous change when the vapour pressure curve is crossed. A large representative selection can be plotted using data from the fundamental EOS available on NIST REFPROP. Table 6.1 lists the key Widom lines studied here using the NIST data, specifying the exact quantity that displays an extremal value when the Widom line is crossed. The functions $z, S_\pm(P), T(P)$ are defined later.

For all fluids, the Widom lines for the different thermodynamic functions commence—by definition—at the exact critical point. Their P,T paths begin following an extrapolation of the vapour pressure curve, but the different Widom lines in a single fluid soon diverge from each other. In addition, a Widom line for a given parameter can take a slightly different path if we find the extremal values as a function of temperature along isobaric paths, instead of the procedure outlined here (finding extremal values as a function of pressure along isothermal paths).

The validity of the Widom lines as a true boundary between liquid-like and gas-like states is therefore debatable. However, when a Widom line is crossed close to the critical point the values of thermodynamic functions do change drastically in a narrow P,T range. Understanding and modelling the Widom lines is therefore undisputably important from the industrial point of view. We will see in Chapter 8 that many industrial

TABLE 6.1 List of Widom lines studied in this work giving the parameter displaying an extremal value and the function fitted as defined below.

Property	Parameter Displaying Extremal Value	Function Fitted
Density ρ or compressibility K^*	$K = \left(\dfrac{\partial \rho}{\partial P} \right)_T$	$z + S_+(P) + V(P)$
Isochoric heat capacity C_V	C_V	$z + S_-(P) + T(P)$
Isobaric heat capacity C_P	C_P	$z + S_-(P) + T(P)$
Sound speed v	v	$z + S_-(P) - T(P)$
Viscosity η	$\left(\dfrac{\partial \eta}{\partial P} \right)_T$	$z + S_-(P) + T(P)$
Thermal conductivity β	$\left(\dfrac{\partial \beta}{\partial P} \right)_T$	$z + S_-(P) + T(P)$

processes, both established and emerging, involve containing fluids close to the critical point where the crossing of a Widom line can drastically alter sample properties.

In the literature, statements on how far the Widom lines extend from the critical point are seemingly contradictory; ranging from $T/T_C \approx 2.5$ and $P/P_C \approx 15$ [6.2] to $T/T_C \approx 5$ and $P/P_C \approx 50$ [6.3][6.4]. This disagreement cannot be justified when, for the most important fluids, the Widom lines can be plotted out accurately using a fundamental EOS backed by large amounts of experimental data collected in different research groups over the decades. As we have discussed, the extremal value constituting a Widom line gradually gets smeared out the further you go from the critical point. Therefore, it is at present a matter of opinion at what point the extremal value becomes sufficiently smeared out to no longer count as a Widom line.

To illustrate this point, Figure 6.2 shows the spike in C_V constituting the Widom line for this parameter in Ar at 151 K (immediately above the critical point) and at successive temperatures up to 200 K, by which the spike has disappeared. But at what point do you make the judgement that there is no longer a sufficiently pronounced extremal value to justify drawing the Widom line? Unfortunately, there is no clear-cut answer to this question. In this chapter we outline a simple phenomenological procedure for fitting the trends in different parameters exhibiting a Widom line (this procedure is universal) and then propose what seem like sensible criteria for deciding where each Widom line ends. There is some commonality between the procedure employed for different Widom lines but unfortunately no universal criterion that will operate for all Widom lines.

Using this methodology, the Widom lines are plotted using the fundamental EOS for key molecular and atomic fluids Ar, H_2O, CH_4, and CO_2 (He and Ne are shown in Appendix A, H_2 and N_2 are shown in Appendix B). The Widom line behaviour in the

---

* Usually compressibility is defined in terms of volume $\left(-\frac{1}{V} \left(\frac{\partial V}{\partial P} \right)_T \right)$. However, in the context of this chapter density is more convenient to work with since volume drops from infinity to a finite value on pressure increase whilst density rises from zero to a finite value.

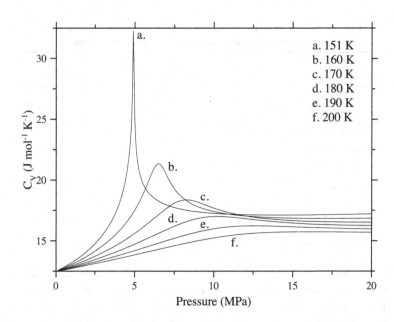

FIGURE 6.2 Extremal value of C_V of Ar marking the Widom line for this parameter shown at a range of temperatures beginning just above the critical point (a.) and extending up to 200 K (f.) at which point the Widom line no longer exists. Data are from NIST REFPROP.

studied fluids is compared and the locus of the Widom lines on the P,T and ρ,T phase diagrams is compared to that of the Frenkel line later in the chapter.

6.1.1 A Simple Phenomenological Fitting Procedure for the Widom Lines

The Widom lines in all fluids can be reproduced from the fundamental EOS accessible via (for instance) NIST REFPROP or the ThermoC software [3.25]. The Widom lines can be plotted simply by reading off the extremal value of the property or its derivative as a function of pressure or temperature. However, for the purposes of the comparisons in this text between the same Widom line in different fluids and different Widom lines for the same fluid, the author has developed a simple function which gives a good fit to Widom line data above ca. $1.1T_C$. The Widom line path is then obtained from this function. It should, however, be understood at the outset that the procedure is entirely phenomenological.

We begin by observing the general trend when a Widom line is crossed, for instance the variation in isochoric heat capacity C_V in CH_4 during an isothermal pressure increase at ca. $1.1T_C$ (Figure 6.3). The heat capacity significantly changes between its value in the low pressure regime and the high pressure regime. This change is represented using a smeared out step function $S(P)$ (equation 6.1). In addition, there is a spike in the heat capacity roughly when the critical isochore is crossed. This is represented using a pseudo-Voigt function $\Gamma(P)$ (equation 6.2). The

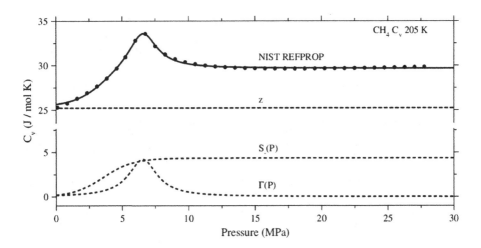

FIGURE 6.3 Trend in C_V for CH_4 at 205 K obtained from NIST REFPROP (circles show selected data points) and fitted using $C_V = z + S_-(P) + \Gamma(P)$. Fit sum is shown with solid line and fit terms are shown with dotted lines.

fit to parameter Π is completed by a constant background value z set to the value of the lowest value datapoint at the low or high pressure end of the range of data considered (equation 6.3).

$$S_\pm(P) = \frac{A}{1 + e^{\frac{B \pm P}{C}}} \qquad (6.1)$$

$$\Gamma(P) = G_h e^{-(\ln 2) \times \left(\frac{P - P_0}{\Delta P}\right)^2} + \frac{L_h}{1 + \left(\frac{P - P_0}{\Delta P}\right)^2} \qquad (6.2)$$

$$\Pi = z + S_\pm(P) \pm \Gamma(P) \qquad (6.3)$$

The form $S_-(P)$ (with positive A, B, C) gives a step function increasing upon pressure increase whilst the form $S_+(P)$ (with positive A, C and negative B) gives a step function increasing upon pressure decrease. Initial values for A, B, C in line with these requirements are given to the regression analysis software (Magicplot Pro) but the fitting process for A, B, C is not constrained in any way.

The Gaussian and Lorentzian components of the pseudo-Voigt function are represented using a common width ΔP but separate peak heights G_h (Gaussian) and L_h (Lorentzian). The common peak width, and additional conditions that $G_h \geq 0$ and $L_h \geq 0$, are constraints that the author has found necessary in previous unrelated work using this function to ensure physically reasonable fits (the peak can otherwise split into two).

The previous section concluded by outlining the problem of deciding where each Widom line should terminate, and the lack of a universal criterion that can be used. Now, we will propose criteria for the termination of each Widom line that we will use for the remainder of this text. In some cases the criteria depend solely on the output (data) from the fundamental EOS, and in some cases use the results of our fitting as outlined in equations 6.1–6.3. For all Widom lines studied, we are forced to make an arbitrary choice of where to cut off the line—but at least it is an arbitrary choice which is applied consistently across different fluids.

We will begin with the Widom line for density/compressibility. In the absence of any phase transitions, the compressibility of all materials (gas, solid, or liquid) generally decreases upon isothermal pressure increase. However, when the Widom line is crossed the compressibility instead rises as the Widom line is approached. We will consider that a Widom line for compressibility exists if, upon isothermal pressure increase, the compressibility increases by at least 1/3rd from its value in the low pressure limit to the maximum value. Figure 6.4 (a) shows the compressibility of fluid Ar at a temperature close to the critical point (155 K, $1.03T_C$), and the temperature at which this criterion is no longer satisfied (240 K, $1.59T_C$).

We will use the same criterion for the heat capacities C_V and C_P. Generally, heat capacity rises upon isothermal pressure increase. This is because the additional interactions

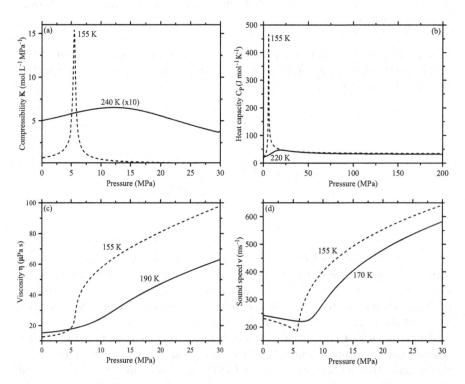

FIGURE 6.4 Trends in compressibility K (a), isobaric heat capacity C_P (b), viscosity η (c), and sound speed v (d) at a temperature close to the commencement of the Widom line (155 K, $1.03T_C$) and the temperature at which the Widom line ceases to exist according to the criteria detailed in the text. Data are for fluid Ar on REFPROP. In the case of the compressibility, the values at 240 K are multiplied by ten for clarity.

between particles in the fluid that take place at higher density allow additional ways for the fluid to store heat, as discussed elsewhere in this text. However, when the Widom line is crossed there is a spike followed by a decrease in heat capacity. This is because, to cause a rise in *temperature* in this P,T region it is necessary to break inter-molecular bonds in an analogous way to the subcritical boiling transition. Hence this spike *followed by a decrease* is the key indicator that the Widom line exists. We will consider that the Widom line exists when this decrease causes the value of the heat capacity to decline by at least 1/3rd from its peak value. Figure 6.4 (b) shows trends in C_P of fluid Ar at 155 K (1.03T_C) and at 220 K (1.46T_C) when this condition is no longer satisfied.

For viscosity and thermal conductivity we will use the results of our fitting procedure to determine the conditions under which we consider the Widom line to exist. Both of these parameters increase steadily in a fluid upon isothermal pressure increase in the absence of any phase transitions. However, they also exhibit a discontinuous increase when the vapour pressure curve is crossed into the liquid state, and a large increase over a narrow P,T range when the Widom line is crossed. The question is, how large is the increase over a narrow P,T range when the Widom line is crossed relative to the value in the low pressure limit? We will consider that a Widom line exists when the increase over a narrow P,T range doubles the parameter's value in the low pressure limit. The size of the increase representing the Widom line will be determined as the area of the peak $\Gamma(P)$ fitted to the derivative (Table 6.1). Figure 6.4 (c) shows example trends in the viscosity in fluid Ar at 155 K (1.03T_C) close to the critical point and at 190 K (1.26T_C) when this criterion is no longer satisfied.

We will finally consider the Widom line in sound speed. The actual change in sound speed when the Widom line is crossed is minimal compared to its absolute value in the high pressure and low pressure limits so the Widom line has a small effect on material properties. As pressure is increased from the low pressure limit, the speed of sound exhibits a small decrease, then increases strongly once the Widom line has been crossed. We will consider the Widom line to exist when the decrease is at least 10% of its value in the low pressure limit. Figure 6.4 (d) shows example trends in sound speed in fluid Ar at 155 K (1.03T_C) close to the critical point and at 170 K (1.13T_C) when this criterion is no longer satisfied.

6.1.2 Some Examples of Widom Line Paths

Using the methodology we have established we can now plot out the paths of Widom lines for some atomic and molecular fluids on the P,T phase diagram. In the appendices Widom lines for more fluids are shown, on both the P,T and ρ,T phase diagrams. Monatomic fluids are shown in Appendix A and molecular fluids are shown in Appendix B. Figure 6.5 shows the Widom lines of Ar, CH_4, H_2O, and CO_2 on the P,T phase diagram. The fitting procedure occasionally struggled to fit the extremely sharp and asymmetric peaks seen in the immediate vicinity of the critical point but generally performed well.

The Widom lines shown in Figure 6.5 are those for a contrasting set of fluids. Ar is the archetypal "model" fluid, composed of electrically neutral spherical particles. The CH_4 molecule is nearly spherical and very weakly polar due to the similarity in

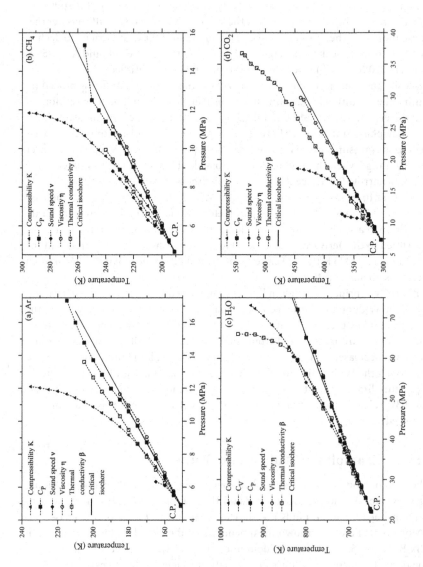

FIGURE 6.5 Widom line paths on the P,T phase diagrams for Ar (a), CH_4 (b), H_2O (c), and CO_2 (d). C.P. marks the critical point in all cases. The anomalously high pressure for the C_P Widom line at 255 K in panel (b) is due to a poor fit upon weakening of the maximum due to temperature increase. Data are from NIST REFPROP, fitted as outlined in the text.

electronegativity between C and H. The CO_2 molecule is linear and also exhibits only weak polarization, whilst H_2O is strongly polarized. Nevertheless, there are clear similarities between the Widom lines in all the studied fluids. In all fluids the Widom lines generally trend to slightly lower density than the critical isochore upon temperature increase, and in all fluids C_P is one of the longest Widom lines whilst the C_v Widom line frequently terminates too close to the critical point to warrant plotting. We therefore, in the next section, collate the Widom line data for a total of eight different fluids to further examine these common features.

6.1.3 The Widom Lines as a Function of Reduced Temperature

Here, and in Appendices A and B, the Widom lines are plotted using the criteria outlined above for a total of eight different fluids. Table 6.2 lists the Widom line termination as a function of reduced temperature $T_R (= T / T_C)$ for these fluids. It is not that meaningful to compare the different Widom lines for the same fluid with the clear exception of the Widom lines for C_V, C_P.

We can see for all fluids examined here, and in Appendices A and B, that when plotted out and terminated according to the same criteria, the Widom line for C_P is far more pronounced and extends much further from the critical point than the Widom line for C_V. Qualitatively, we may understand this as follows. The transition across the Widom line is analogous to the subcritical boiling transition, in which bonds such as van der Waals bonds or hydrogen bonds between the particles in the fluid are broken, which can only happen if the volume increases. When we increase temperature at constant pressure (and therefore increasing volume) this process can take place; when we increase temperature at constant volume it cannot. Hence the Widom line for C_V is much weaker than the Widom line for C_P.

TABLE 6.2   Widom line termination reduced temperatures T_R for fluids studied here and in Appendices A and B. (a) indicates that the criteria outlined in this section for the termination of the Widom lines could not be used due to the non-standard nature of the Widom line. The Widom line termination points for He were not included in the averages due to fluid He not satisfying the classical fluid conditions in the ρ, T conditions where the Widom lines lie.

| Fluid | Widom Line Termination Reduced Temperature T_R | | | | | | |
	κ	C_V	C_P	v	η	β	Average
He	1.9	-	4.4	1.3	1.9 (a)	2.4 (a)	2.4
Ne	1.5	-	1.5	1.1	1.1	1.2	1.3
Ar	1.6	-	1.4	1.1	1.2	1.4	1.3
H_2	1.7	1.1	2.4	1.1	1.8	1.5	1.6
N_2	1.5	-	1.3	1.2	1.5	1.5	1.4
CH_4	1.6	-	1.3	1.2	1.2	1.3	1.3
H_2O	1.4	-	1.3	1.3	1.2	Not fittable	1.3
CO_2	1.5	-	1.3	1.2	1.4	1.8	1.4
Average	1.5	-	1.5	1.2	1.3	1.2	

In addition, it is not meaningful to study the absolute values for the reduced temperature range T_R covered by the Widom lines. We could extend or reduce this temperature range as we wished simply by changing the criteria used to define that a Widom line is present.

What is meaningful, however, is to compare the Widom lines for different substances. The clear conclusion from such a comparison is that the Widom lines for all the fluids studied behave in a similar manner as a function of the reduced variables T_R, P_R despite the differing nature of the fluids studied. The length of the Widom lines in this case is roughly the same for all fluids studied, and it is generally the same Widom lines for all fluids that extend the furthest from the critical point. For instance, the compressibility Widom line is always one of those that extends the furthest and trends to the lowest density relative to the critical isochore. In all cases, the Widom lines by $T_R = 2$ no longer mark a significant and narrow transition in fluid properties.

The observation of these trends is unlikely to be a consequence of the fact that in all cases we are fitting to the fundamental EOS for the fluid in question, i.e., of some feature in the fundamental EOS. This is because the fundamental EOS (a) reproduces the vast amount of experimental data with extremely good accuracy and (b) is different for each fluid studied (at least as far as the residual component of the reduced Helmholtz function is concerned). These trends are also in agreement with theoretical findings on model fluids [6.1][6.5][6.6] and with what is expected from the principle of corresponding states.

6.1.4 The Widom Lines in Relation to the Vapour Pressure Curve

The Widom lines were introduced at the beginning of this chapter as the thermodynamic continuation of the vapour pressure curve. This is correct, however the Widom lines are also fundamentally different from the vapour pressure curve in several respects. Firstly, there is no longer a first order (or second order) phase transition so the Widom lines cannot be plotted out using the Clausius-Clapeyron equation or the equivalent for a second order transition. Secondly, as we have seen, the different Widom lines for a given sample diverge from each other rapidly upon P, T increase.

Thirdly, the Widom line transition (as opposed to the boiling/condensation transition across the vapour pressure curve) takes place when the amount of thermal energy available exceeds a critical value. In any fluid, there is a certain interaction energy ε between particles; the depth of the potential well described using the pair potential $\mathcal{V}(r)$ (for instance the Lennard-Jones potential shown in Figure 2.1). The exact nature of the potential $\mathcal{V}(r)$ between particles in a fluid will vary depending on the sample, but it will always share the general features of the Lennard-Jones potential that are listed and explained in Chapter 2 (equations 2.5–2.7).

As discussed extensively in other texts [1.1][1.16][2.2][2.4] and in Section 2.4, the most fundamental definition of the critical temperature is that it marks the temperature where $k_B T \approx \varepsilon$, though this relation is very approximate. That is, it is the temperature where thermal agitation provides sufficient energy to overcome the attraction between particles in the fluid. Therefore, whilst the Widom lines are the thermodynamic continuation of the vapour pressure curve they cannot be viewed as thermodynamically equivalent or similar to the vapour pressure curve.

6.1.5 The Widom Lines as a Function of Density

Thus far, we have plotted the Widom lines solely as a function of pressure. However, plotting them also as a function of density is a useful aid to understanding. When plotted against density, the Widom line transitions are far more gradual. Figure 6.6 shows two Widom line transitions in Ar at 180 K ($1.2T_C$) plotted as a function of pressure and of density. The C_P transition is far more pronounced as a function of pressure than density, and the viscosity transition has completely ceased to exist when plotted as a function of density. The inset to Figure 6.6 (b) shows the ρ,P EOS at 180 K.

On this basis, it is tempting to begin with the Widom line for density/compressibility and understand other Widom lines such as that for viscosity as consequences of the Widom line for density. On the other hand, the divergence of the different Widom lines would still remain when they are plotted as a function of density and (as shown in Figure 6.6 (a)) there is still a peak in some parameters when the line is plotted as a function of density.

6.2 The Fisher-Widom Line

We have considered the radial distribution function $g(r)$ of fluids at various points in the text thus far and seen very different behaviour depending on the ρ,T conditions. In the low density limit (Figure 4.8) the function decays exponentially following a single peak. On the other hand, at higher density there are several subsequent peaks in the function at higher r, with exponentially damped amplitude, indicating long-range order (for instance in Figure 4.7 (b)). The Fisher-Widom line marks the transition on the P,T or ρ,T phase diagram between the regions there the oscillatory decay in $g(r)$ is present or not.

Plotting out the path of the Fisher-Widom line on the phase diagram using experimental data on a real fluid would require an unfeasibly large amount of X-ray or neutron diffraction beamtime at a central facility; to the author's knowledge the P,T path of the Fisher-Widom line has therefore not been measured experimentally. It has, however, been plotted theoretically for some model fluids in one dimension [6.7] and later in three dimensions [2.4][6.8]. Figure 6.7 plots the Fisher-Widom line for the Lennard-Jones fluid [2.4][6.9] alongside the vapour pressure curve as a function of the reduced variables P_R, T_R. To obtain a Fisher-Widom line it is necessary to truncate the pair potential in use. As seen, the Fisher-Widom line commences just below the critical temperature then trends to higher pressure upon temperature increase. The line trends to slightly lower density upon temperature increase, as labelled.

It is not presently clear how far beyond the critical temperature the Fisher-Widom line should extend. Inspection of real diffraction data collected at significantly supercritical temperature [1.7][4.8] suggests that the transition from exponential to damped oscillatory decay in $g(r)$ at high r does still exist, but becomes significantly smeared out.

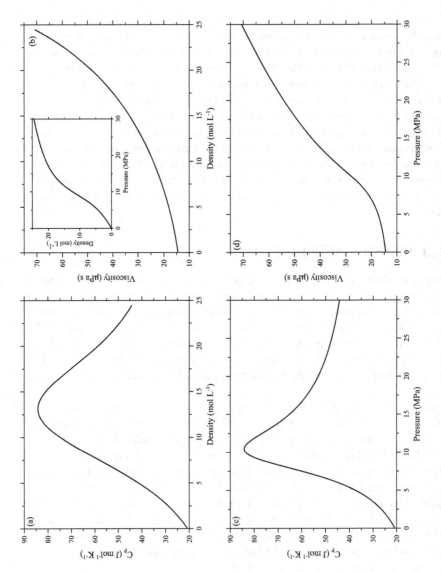

FIGURE 6.6 Widom lines for C_P and viscosity in Ar at 180 K ($1.2T_C$), plotted as a function of density ((a) and (b)) and of pressure ((c) and (d)). The inset to panel (b) shows the ρ, P EOS of Ar at 180 K. Data are from NIST REFPROP.

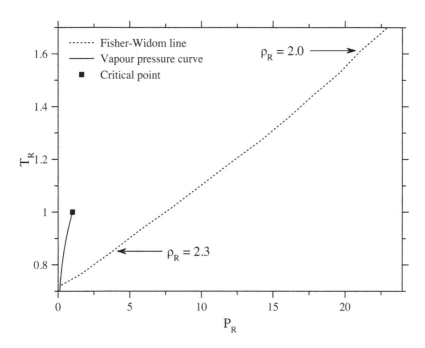

FIGURE 6.7 Fisher-Widom line and vapour pressure curve for the Lennard-Jones fluid. Data are compiled using tabulated simulation data and P,V,T EOS from refs. [2.4][6.9] and have been smoothed. Labels ρ_R indicate reduced density at labelled points along the Fisher-Widom line.

6.3 The Joule-Thomson Inversion Curve

The Joule-Thomson effect is the change of temperature of a fluid during an adiabatic expansion through a porous plug/valve from P_1,T_1 to P_2,T_2 (see Figure 6.8 (b)). The expansion can be considered as a continuous flow process with the fluid starting in equilibrium at P_1,T_1 and ending in equilibrium at P_2,T_2. The Joule-Thomson line (or inversion curve) divides the areas on the phase diagram in which such an adiabatic expansion with an infinitesimally small pressure change warms, or cools, the fluid. There exists a whole family of different inversion curves corresponding to different parameters remaining constant during the expansion. The Joule-Thomson curve is the most important due to the applications in refrigeration and liquefaction of gases so will be described first, followed by the Boyle curve and Amagat curve in Section 6.4. The Joule-Thomson effect is also known as the Joule-Kelvin effect and the Joule-Thomson curve is also known as the Charles curve.

We begin our treatment of the Joule-Thomson effect by considering the work W done by one mole of the fluid whilst it passes through the plug:

$$W = P_2 V_2 - P_1 V_1 \tag{6.4}$$

Since the expansion is adiabatic, this work must be done at the expense of the internal energy U in the system. Therefore, we may write $U_1 - U_2 = P_2 V_2 - P_1 V_1$ and conclude that

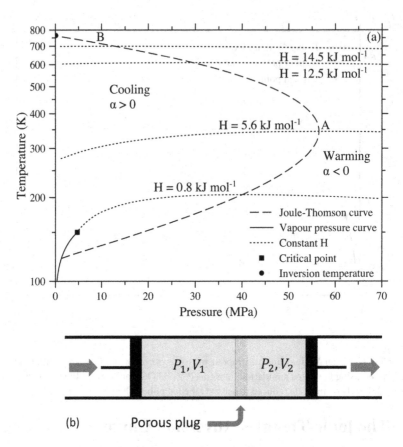

FIGURE 6.8 (a) Joule-Thomson inversion curve (dashed line) and vapour pressure curve (solid line) for Ar. The Joule-Thomson inversion curve joins the temperature maxima of lines of constant enthalpy (dotted lines). (b) Schematic of Joule-Thomson expansion process. Data are plotted from the fundamental EOS [4.1] using NIST REFPROP and the ThermoC code [3.25].

the process is isenthalpic, as expected. Therefore, the rate of change of temperature in the process is described with the following expression. This is for the Joule-Thomson coefficient α.

$$\alpha = \left(\frac{\partial T}{\partial P} \right)_H \tag{6.5}$$

We would like to obtain an expression for α in terms of quantities that can be directly measured and modelled (heat capacity and P,V,T EOS). We do this by considering H as a function of U,P,V and setting $\delta H = 0$ for the process:

$$\delta H = \delta(U + PV) = \delta U + P\delta V + V\delta P = 0$$

Using the standard identity $\delta U = T\delta S - P\delta V$, we obtain:

$$\delta H = T\delta S + V\delta P = 0 \tag{6.6}$$

We can describe δS in terms of the variables P,T:

$$\delta S = \left(\frac{\partial S}{\partial T}\right)_P \delta T + \left(\frac{\partial S}{\partial P}\right)_T \delta P$$

Then we recall the definition of the heat capacity at constant pressure C_P in terms of S, set $\delta H = 0$, and utilize one of the Maxwell relations to obtain the final expression for the Joule-Thomson coefficient α in terms of C_P and parameters obtainable from the PVT EOS.

$$C_P = T\left(\frac{\partial S}{\partial T}\right)_P$$

$$\left(\frac{\partial S}{\partial P}\right)_T = -\left(\frac{\partial V}{\partial T}\right)_P$$

$$\alpha = \left(\frac{\partial T}{\partial P}\right)_H = \frac{1}{C_P}\left[T\left(\frac{\partial V}{\partial T}\right)_P - V\right] \tag{6.7}$$

Since any Joule-Thomson expansion process is isenthalpic we plot lines of constant enthalpy H on the P,T phase diagram (Figure 6.8) and see that P,T space is divided into areas where a Joule-Thomson expansion warms the fluid ($\alpha < 0$) or cools the fluid ($\alpha > 0$). The curves in Figure 6.8 are the Joule-Thomson and vapour pressure curves for Ar, plotted using the ThermoC software [3.25] and fundamental EOS [4.1] available on NIST REFPROP. The lines of constant enthalpy at higher temperature terminate at supercritical temperature and zero pressure, however the lower temperature lines ($H < 0.8$ kJ mol^{-1}) are terminated by the vapour pressure curve. The 0.8 kJ mol^{-1} line terminates almost exactly at the critical point, with the liquid state having lower enthalpy than this and the gas state having higher enthalpy.

The inversion curve, by definition, joins the maxima of the lines of constant enthalpy H. It intersects $P = 0$ at a specific temperature called the inversion temperature. Only below this temperature is it possible to cool a fluid using a Joule-Thomson expansion. For example, to cool He using a Joule-Thomson expansion it is necessary to first precool using some other process to get below the inversion temperature of 45 K. In addition, we can observe that only above the inversion temperature is truly ideal gas behaviour realised (i.e., the Joule-Thomson effect ceases to exist due to the isenthalpic process becoming isothermal).

The Joule-Thomson curve always encircles the critical point, i.e., at the critical point $\alpha > 0$) and we can prove that this is the case for the simple example of a van der Waals fluid. We begin by differentiating the van der Waals equation in the form $T(P,V)$ (equation 1.1, reproduced below as equation 6.8):

$$RT = \left(P + \frac{a}{V^2}\right)(V - b) \tag{6.8}$$

$$R\left(\frac{\partial T}{\partial V}\right)_P = (V-b)\left[\frac{RT}{(V-b)^2} - \frac{2a}{V^3}\right] \tag{6.9}$$

We then utilise the van der Waals equation in form $P(V,T)$ (equation 2.14) to obtain:

$$\left(\frac{\partial P}{\partial V}\right)_T = -\frac{RT}{(V-b)^2} + \frac{2a}{V^3}$$

$$R\left(\frac{\partial T}{\partial V}\right)_P = -(V-b)\left(\frac{\partial P}{\partial V}\right)_T$$

$$\left(\frac{\partial V}{\partial T}\right)_P = -\frac{R}{V-b}\left(\frac{\partial V}{\partial P}\right)_T \tag{6.10}$$

This allows us to express α (cf. equation 6.7) in a form that we can easily understand using the nature of the isotherms at the critical point:

$$\alpha = -\frac{1}{C_P}\left[\frac{RT}{V-b}\left(\frac{\partial V}{\partial P}\right)_T + V\right] \tag{6.11}$$

Or, in reduced pressure, temperature, and volume utilising equations 2.16–2.20:

$$\alpha = -\frac{V_C}{C_P}\left[\frac{8T_R}{3V_R-1}\left(\frac{\partial V_R}{\partial P_R}\right)_T + V_R\right] \tag{6.12}$$

As the critical point is approached $V_R \rightarrow 1$ etc. and therefore:

$$\alpha \rightarrow -\frac{V_C}{C_P}\left[4\left(\frac{\partial V_R}{\partial P_R}\right)_T + 1\right] \tag{6.13}$$

Since $\left(\frac{\partial V_R}{\partial P_R}\right)_T \rightarrow -\infty$, at the critical point $\alpha > 0$ always. The spike in C_P at the critical point (behaviour similar to that for C_V shown in Figure 2.4) ensures that the value of α remains finite. An analytical expression for the entire P,T path of the Joule-Thomson curve has been obtained for the van der Waals EOS [6.10] (given below). Unfortunately, analytical expressions for the path of the Joule-Thomson curve cannot in practice be obtained from EOS that are quantitatively accurate.

$$P = -\frac{(bRT-2a)\left(2abRT + 5bRT\sqrt{2abRT} + 3(bRT)^2 - 2a\sqrt{2abRT}\right)a}{b^2\left(\sqrt{2abRT} + bRT\right)\left(2a + \sqrt{2abRT}\right)^2}$$

It can be demonstrated [6.11] that the inversion curve can be described in terms of the competition between the effects of attractive and repulsive forces between particles. On the $\alpha > 0$ side attractive forces dominate and on the $\alpha < 0$ side repulsive forces dominate. Whilst correct, this statement is open to misinterpretation. In the region marked A in Figure 6.8, repulsive forces dominate since the particles are forced too close together for attractive forces to have any effect. In region B, repulsive forces dominate because sufficient thermal energy is available to overcome the attractive forces, but not to overcome the much stronger repulsive forces when particles collide.

6.4 A General Approach to Inversion Curves

We have commenced our discussion of inversion curves with the Joule-Thomson curve because it is by far the most important due to its physical interpretation and applications in refrigeration. There are, however, other inversion curves. In addition to the Joule-Thomson curve there is the Boyle curve, the Amagat curve, and the Zeno curve. In this section a unified approach to understanding all of these curves is shown, following the methodology of Brown [3.27].

The inversion curves listed above exist for all condensing fluids and can be understood in terms of the compression factor. Figure 6.9 shows a contour plot of the

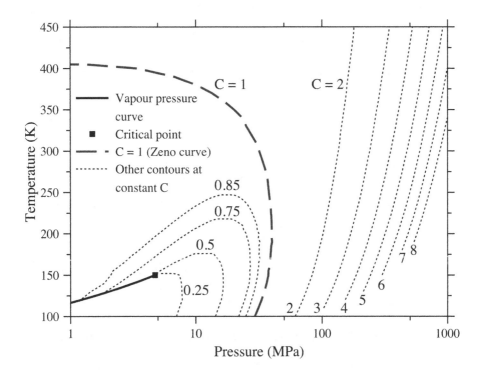

FIGURE 6.9 Contour diagram showing lines of constant compression factor C for fluid Ar, taken from NIST REFPROP. The Zeno curve ($C=1$) is shown in bold. Lines are plotted up to the highest density at which the fundamental EOS is backed by experimental data.

compression factor $C = PV / RT$ for a typical condensing fluid (Ar), covering the gas, liquid, and supercritical fluid regions. As we can see, there exists a line in the phase diagram linking the points where $C=1$. This line is the Zeno curve, also known as the Batschinski curve. It corresponds to the case where the compression factor is that of an ideal gas and is the only *zeroth order* characteristic curve. The Zeno curve condition $C=1$ is also met at all temperatures in the low pressure limit. The endpoint of the Zeno curve on the temperature axis is the point where the low pressure limit is approached along an isotherm satisfying the condition $\left(\frac{\partial C}{\partial P}\right)_T = 0$. This condition is satisfied at the Boyle temperature.

6.4.1 First Order Inversion Curves: Definitions

We show in this section that there are exactly three possible first order inversion curves that cannot intersect with each other, and that one of these curves shares an endpoint with the Zeno curve on the temperature axis. Given the variables that determine C for the condensing fluid, it is possible to define three *first order* inversion curves (equations 6.14–6.16):

$$\left(\frac{\partial C}{\partial T}\right)_P = 0 \tag{6.14}$$

$$\left(\frac{\partial C}{\partial T}\right)_V = 0 \tag{6.15}$$

$$\left(\frac{\partial C}{\partial P}\right)_T = 0 \tag{6.16}$$

The curve of $\left(\frac{\partial C}{\partial V}\right)_T = 0$ would lie in the same location as that given by equation 6.16 so need not be considered separately. Considering $C(P,T)$ or $C(V,T)$ we obtain the following proof that the three first order inversion curves cannot intersect:

$$\delta C = \left(\frac{\partial C}{\partial T}\right)_P \delta T + \left(\frac{\partial C}{\partial P}\right)_T \delta P = \left(\frac{\partial C}{\partial T}\right)_V \delta T + \left(\frac{\partial C}{\partial V}\right)_T \delta V$$

From the above we can see that if any two of the conditions 6.14–6.16 were to be satisfied simultaneously the third would also have to be satisfied and $\delta C = 0$, i.e., there is a saddle point or absolute maximum/minimum in C. There is no known mechanism for a fluid to exhibit such a property.

We can demonstrate easily that equation 6.14 is equivalent to the Joule-Thomson coefficient being zero. From equation 6.7 this latter condition is satisfied when:

$$T\left(\frac{\partial V}{\partial T}\right)_P = V \tag{6.17}$$

Meanwhile a calculation of $\left(\frac{\partial C}{\partial T}\right)_P$ yields, for any fluid:

$$R\left(\frac{\partial C}{\partial T}\right)_P = -\frac{PV}{T^2} + \frac{P}{T}\left(\frac{\partial V}{\partial T}\right)_P$$

Substituting equation 6.17 into here yields $\left(\frac{\partial C}{\partial T}\right)_P = 0$ so condition 6.14 is equivalent to the condition for the Joule-Thomson curve.

The condition in equation 6.15 is equivalent to the Amagat inversion curve, linking temperature maxima of lines of constant internal energy and defined by equation 6.18:

$$\left(\frac{\partial T}{\partial P}\right)_U = 0 \qquad (6.18)$$

From equation 6.15 we obtain, for any fluid:

$$R\left(\frac{\partial C}{\partial T}\right)_V = -\frac{PV}{T^2} + \frac{V}{T}\left(\frac{\partial P}{\partial T}\right)_V = 0$$

Hence equation 6.15 is equivalent to:

$$\left(\frac{\partial P}{\partial T}\right)_V = \frac{P}{T} \qquad (6.19)$$

The internal energy can be constrained using equation 6.19 and a standard thermodynamic identity[†][2.8]. The constraint on the internal energy is:

$$\left(\frac{\partial U}{\partial V}\right)_T = 0 \qquad (6.20)$$

Calculating $\delta U(V,T)$ in the usual manner we obtain:

$$\delta U = \left(\frac{\partial U}{\partial T}\right)_V \delta T + \left(\frac{\partial U}{\partial V}\right)_T \delta V$$

Since we are concerned with the case of $\delta U = 0$ we obtain:

$$\left(\frac{\partial U}{\partial T}\right)_V \left(\frac{\partial T}{\partial P}\right)_U + \left(\frac{\partial U}{\partial V}\right)_T \left(\frac{\partial V}{\partial P}\right)_U = 0$$

Recalling that $\left(\frac{\partial U}{\partial T}\right)_V = C_V$ so is always finite, we apply equation 6.20 and obtain equation 6.18. Our method has shown that, like the Joule-Thomson curve, the Amagat curve can indicate a boundary between gas-like and liquid-like behaviour to a limited extent.

[†] $P + \left(\frac{\partial U}{\partial V}\right)_T = T\left(\frac{\partial P}{\partial T}\right)_V$

Equation 6.20 shows that the Amagat curve marks a boundary between a region on the phase diagram where attractive forces dominate (so compression decreases the internal energy) and where repulsive forces dominate (so compression increases the internal energy).

We deal finally with equation 6.16. We show that this is equivalent to the Boyle inversion curve, linking temperature maxima of lines of constant (PV), so defined by equation 6.21:

$$\left(\frac{\partial T}{\partial P} \right)_{PV} = 0 \tag{6.21}$$

From equation 6.16 we obtain:

$$R\left(\frac{\partial C}{\partial P} \right)_T = \frac{V}{T} + \frac{P}{T}\left(\frac{\partial V}{\partial P} \right)_T = 0$$

So the condition 6.16 is equivalent to:

$$\left(\frac{\partial V}{\partial P} \right)_T = -\frac{V}{P} \tag{6.22}$$

We can demonstrate the equivalence of this condition to equation 6.21 by considering a change δV in the usual manner:

$$\delta V = \left(\frac{\partial V}{\partial T} \right)_P \delta T + \left(\frac{\partial V}{\partial P} \right)_T \delta P$$

Consideration of this condition at constant (PV) yields:

$$\left(\frac{\partial V}{\partial P} \right)_{PV} = \left(\frac{\partial V}{\partial T} \right)_P \left(\frac{\partial T}{\partial P} \right)_{PV} + \left(\frac{\partial V}{\partial T} \right)_P$$

It is easy to show that $\left(\frac{\partial V}{\partial P} \right)_{PV} = -\frac{V}{P}$ for any fluid, by setting $\delta(PV) = 0$. Applying this and the condition in equation 6.21 to the above equation yields an equation identical to equation 6.22. Equation 6.16 is therefore the condition to be satisfied on the Boyle inversion curve, equivalent to equation 6.21. Equation 6.16 is also the condition satisfied by the Zeno curve on the temperature axis, so the two curves must share a common endpoint here. The full nomenclature and defining characteristics utilized in the text of the zeroth order inversion curve and the three first order inversion curves are summarized in Table 6.3 below. Comparing the conditions for the different curves, we can observe that the Boyle curve, Joule-Thomson curve, and Amagat curve are temperature maxima of lines of constant (PV), constant H, and constant U respectively on the phase diagram.

Some other equivalent mathematical conditions are given in refs. [6.12][6.13][6.14]. Figure 6.10 shows the zeroth order and all first order inversion curves in their entirety for fluid Ar. We will now discuss in more detail the path of the curves on the P,T phase diagram.

TABLE 6.3 Nomenclature and defining characteristics of the zeroth order and first order inversion curves for a condensing fluid. The definitions utilizing the residual part of the reduced Helmholtz function have been written using the notation from Chapter 3 of this text.

Name	Principal Mathematical Condition	Mathematical Condition in Terms of the Residual Reduced Helmholtz Function [6.14].	Other Mathematical Conditions Utilized in the Text
Zeno curve (Batschinski curve)	$C = 1$	$\left(\dfrac{\partial F_R'}{\partial \rho'}\right)_\mu = 0$	
Boyle inversion curve	$\left(\dfrac{\partial T}{\partial P}\right)_{PV} = 0$	$\left(\dfrac{\partial F_R'}{\partial \rho'}\right)_\mu + \rho'\left(\dfrac{\partial^2 F_R'}{\partial \rho'^2}\right)_\mu = 0$	$\left(\dfrac{\partial C}{\partial P}\right)_T = 0, \left(\dfrac{\partial C}{\partial V}\right)_T = 0$
Joule-Thomson inversion curve (Charles curve)	$\left(\dfrac{\partial T}{\partial P}\right)_H = 0$	$\left(\dfrac{\partial F_R'}{\partial \rho'}\right)_\mu + \rho'\left(\dfrac{\partial^2 F_R'}{\partial \rho'^2}\right)_\mu + \mu\left(\dfrac{\partial^2 F_R'}{\partial \rho' \partial \mu}\right) = 0$	$\left(\dfrac{\partial C}{\partial T}\right)_P = 0, \left(\dfrac{\partial V}{\partial T}\right)_P = \dfrac{V}{T}$
Amagat inversion curve (Joule curve)	$\left(\dfrac{\partial T}{\partial P}\right)_U = 0$	$\left(\dfrac{\partial^2 F_R'}{\partial \rho' \partial \mu}\right) = 0$	$\left(\dfrac{\partial C}{\partial T}\right)_V = 0$

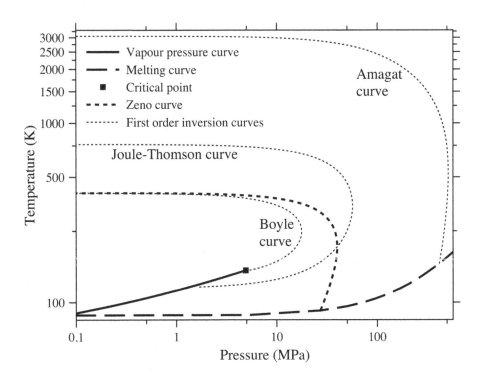

FIGURE 6.10 Zeroth order (Zeno curve) and first order (Amagat curve, Joule-Thomson curve, and Boyle curve) inversion curves for fluid Ar. Fluid data are obtained from the fundamental EOS [4.1] using the ThermoC code [3.25], and melting curve data are obtained from Abramson [4.53].

6.4.2 First Order Inversion Curves: Path on the P,T Phase Diagram

For all fluids the path of the first order inversion curves follows certain characteristics; we have already proved that they cannot intersect. In addition, they all have endpoints along the temperature axis and approach these endpoints with zero gradient in the P,T plane.

We can prove that the Boyle curve intersects the temperature axis at the lowest temperature, followed by the Joule-Thomson curve and finally the Amagat curve at the highest temperature. Since we proved earlier that none of the curves can intersect, it follows that the Amagat curve completely encloses the Joule-Thomson curve, which in turn completely encloses the Boyle curve.

To form this proof, we need to use the mathematical conditions given in Table 6.3 to define the endpoint of each curve in terms of the second virial coefficient. In the low pressure limit we can utilize the virial equation using only the first two terms (cf. equation 1.2):

$$C = 1 + \frac{B_2(T)}{V} \tag{6.23}$$

Beginning with the Boyle curve, we utilize the condition from Table 6.3, $\left(\frac{\partial C}{\partial V}\right)_T = 0$. Applied to equation 6.23 this yields $B_2 = 0$ at the intersection of the Boyle curve with the temperature axis.

Applying the condition for C along the Joule-Thomson curve from Table 6.3 to equation 6.23 we obtain:

$$\left(\frac{\partial C}{\partial T}\right)_P = \frac{1}{V}\frac{dB_2}{dT} - \frac{B_2}{V^2}\left(\frac{\partial V}{\partial T}\right)_P = 0 \tag{6.24}$$

Applying equation 6.17 which is satisfied throughout the Joule-Thomson curve gives us the condition met at the intersection of the Joule-Thomson curve with the temperature axis:

$$\frac{dB_2}{dT} = \frac{B_2}{T}$$

Applying the Amagat curve condition on C (equation 6.15) to equation 6.23 leads to the condition for the termination of the Amagat curve:

$$\frac{dB_2}{dT} = 0$$

The coefficient B_2 is determined solely by the form of the pair potential $\mathcal{V}(r)$ (equation 1.3). In Figure 6.11 we plot $B_2(T)$ as a function of temperature for the Lennard-Jones potential (equation 2.4). We use the arbitrary but physically realistic parameters $a = 1$ Å and $\varepsilon = 50k_B$ (corresponding to $T_C \sim 50$ K). The conditions given above for the termination of the first order curves along the temperature axis are annotated. Note that all the curves terminate at $T > T_C$.

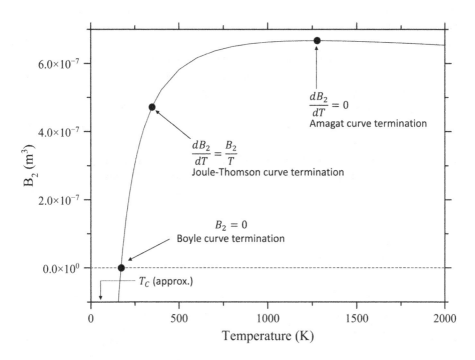

FIGURE 6.11 Plot of the second virial coefficient of a Lennard-Jones fluid as a function of temperature (parameters given in text). Annotations mark criteria and temperatures for termination of the first order characteristic curves on the temperature axis and the critical temperature T_C.

Furthermore, we can prove that the zeroth and first order inversion curves must intersect the temperature axis with zero gradient in the P,T plane. Since the virial coefficients are independent of P,V the temperatures satisfying the conditions given in Figure 6.11 are also independent of P,V. Thus the Amagat, Joule-Thomson, and Boyle curves have zero gradient until density has increased to the point that terms dependent on $B_3(T)$, representing three-body interactions, must also be included in the EOS. The Zeno curve has zero gradient at the intersection with the temperature axis also since it approaches the temperature axis at the (constant) temperature satisfying $B_2(T)=0$, the same as the Boyle curve.

The trend for $B_2(T)$ shown in Figure 6.11 is determined by the mathematical form of the pair potential $V(r)$. Provided that $V(r)$ is repulsive at the smallest r, attractive at moderate r, and $\rightarrow 0$ as $r \rightarrow \infty$ (conditions summarized in equations 2.5–2.7, always true for a condensing fluid), $B_2(T)$ has the form shown in Figure 6.11 and the first order characteristic curves terminate in the order shown.

Brown [3.27] also discussed the low temperature termination of the zeroth and first order inversion curves, proposing that all the curves should continue to subcritical temperatures at supercritical pressures, then trend to lower P,T. The Zeno curve should terminate on the liquid side of the vapour pressure curve at the low pressure, low temperature limit. The first order inversion curves should terminate at points along the vapour pressure curve consistent with their paths not crossing—the Boyle curve terminating very close to the critical point.

As shown above, the Boyle curve does terminate on the vapour pressure curve close to the critical point, followed by the Joule-Thomson curve at lower P,T. However, as shown here and pointed out by other authors [6.12], the Amagat curve and Zeno curve are terminated in practice by intercepting the melting curve. We shall therefore not discuss the low P,T termination of the inversion curves further in this text.

6.4.3 Zeroth and First Order Inversion Curves: Can We Measure Them? Do We Need to Measure Them?

In this section we will discuss the justification for studying the different inversion curves, what physical conditions are described by the curves in real life and the experimental evidence for their existence.

The short answer to the question posed in the section title "Can we measure them?" is yes. As shown in Table 6.3, at least one of the mathematical conditions for each curve is a criterion in terms of the EOS variables P,V,T only. Thus, if the EOS is known then the zeroth and first order characteristic curves can be determined. Whether a given analytic EOS actually yields physically realistic inversion curves with the properties that we have derived from first principles is another matter entirely; often this is not the case. The ability to plot physically realistic inversion curves (i.e., fulfilling the criteria that the Boyle curve is enclosed by the Joule-Thomson curve, enclosed by the Amagat curve) has therefore been proposed as a stringent test of how well an EOS describes the real fluid (refs. [3.4][6.12] and Section 6.4.4). Many commonly used EOS cannot generate all of these curves correctly. In particular, no cubic EOS is able to generate the Amagat curve correctly (see further discussion in ref. [6.13]).

Whilst it is possible to plot the curves, do they correspond to changes in physically important properties? We will begin our discussion with the Zeno curve. The author argues that the Zeno curve is of limited relevance. It is marked by the criterion $C=1$ but the compression factor C is not a physically meaningful property in its own right. $C=1$ may be true throughout the curve but the physical reasons why $C=1$ as the curve approaches the liquid side of the vapour pressure curve in the low temperature limit are completely different from the reasons $C=1$ as the curve approaches the temperature axis at the Boyle temperature. On the temperature axis at the Boyle temperature $C=1$ because the sample is behaving as an ideal gas with no repulsive or attractive interaction between the particles. At the high pressure turning point of the Zeno curve $C=1$ due to the trade-off between strong attractive forces between particles and strong repulsive forces between particles and as the low temperature limit is approached $C=1$ due to the attractive forces between particles playing a dominant role.

If we wish to measure the Zeno curve, it may be easily plotted directly from a suitable EOS backed by experimental data. Even the van der Waals EOS results in a physically reasonable Zeno curve (which is linear on the ρ,T phase diagram [6.15]). Figure 6.12 shows the Zeno curve (calculated using the ThermoC code [3.25]) plotted in relation to the experimentally observed vapour pressure and melting curves of CH_4 and H_2O (Ar was shown in Figure 6.10). In all cases we observe that, as expected, the Zeno curve

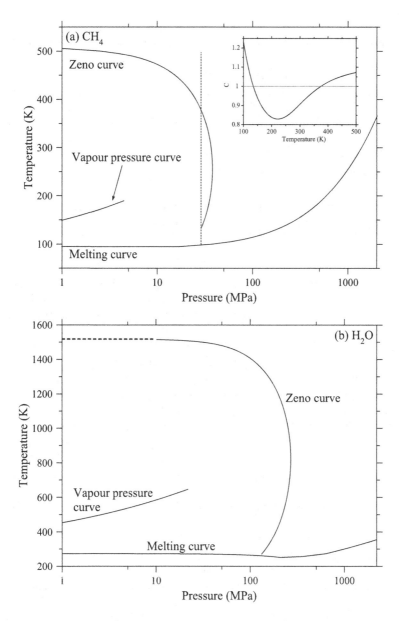

FIGURE 6.12 Experimental vapour pressure, melting, and Zeno curves for CH_4 (a) and H_2O (b) obtained from references [3.16][3.25][4.53][6.16][6.17]. The vapour pressure and Zeno curves are calculated from these fundamental EOS using the ThermoC code [3.25]. The inset in (a) shows C as a function of temperature along the dotted isobar aligning with the experimentally observed endpoint of the Zeno curve. The dotted line in (b) indicates the expected continuation of the Zeno curve to the temperature axis. An accurate calculation of the Zeno curve in this region is not possible due to C being very weakly dependent on P,T in this region (in the low pressure limit, $C=1$ for all T).

is terminated by the melting curve at low temperature. In the cases of Ar and H_2O the Zeno curve almost meets the melting curve, whilst in the case of CH_4 the compression factor (see inset) rises above 1 at significantly higher temperature than the melting curve.

The first order curves may also be plotted directly from a suitable EOS, for instance the curves for Ar plotted from the fundamental EOS for this fluid in Figure 6.10. The physically relevant condition defining each curve is whether an expansion warms or cools the fluid, under different conditions (Table 6.3)—constant (PV) (Boyle), constant U (Amagat), or constant H (Joule-Thomson). The constant H condition corresponds to an adiabatic expansion. Thus, the Joule-Thomson curve is the only inversion curve with widespread industrial relevance.

6.4.4 Use of Zeroth Order and First Order Inversion Curves to Verify Equations of State

Continuing from our discussion at the end of Chapter 3, the inversion curves described provide a useful tool to verify the performance of different fluid EOS or, alternately, to verify the performance of the software used to plot the EOS. Their principal relevance is to the extrapolation of EOS beyond the P,T range in which they are backed by experimental data. Generally, in physics and chemistry we use mathematical methods based closely on first principles, permitting (careful) extrapolation. However, as we saw in Chapter 3, due to the complexity of the liquid and supercritical fluid states there are no EOS performing well over a wide P,T range that are obtained directly from first principles; all are complex semi-empirical expressions. Therefore, it cannot be generally assumed that a fluid EOS will provide accurate information when extrapolated beyond the P,T range in which it is backed by experimental data.

The inversion curves described extend to extremely high pressure and temperature; often far beyond the conditions for which good data are available. In addition, their position is extremely sensitive to a very small change in the P,V,T variables. Therefore, the ability of an EOS to predict physically reasonable inversion curves outside the P,T range in which it is backed by experimental data is a useful test of how safe it is to extrapolate the EOS. In Table 6.3 we saw that all the inversion curves can be directly defined in terms of the P,V,T EOS only, and in terms of derivatives of the Helmholtz function for use with the fundamental EOS.

The only data shown thus far in this chapter for which significant extrapolation of an EOS was required are the Amagat curve data for Ar shown in Figure 6.10 and the Zeno curve for H_2O shown in Figure 6.12. To produce the Ar Amagat curve for instance, the fundamental EOS was extrapolated from 700 K all the way up to 3000 K. In this case, the Amagat curve produced has a physically reasonable path, but this is not always the case. Span and Wagner [6.14] showed an example where extrapolation of two different EOS for CO_2 produced densities at the high pressure end of the Amagat curve differing by just 1%, yet this minute discrepancy caused one EOS to produce an Amagat curve with a completely unphysical path at the highest pressures.

Figure 6.13 (a) shows another example to illustrate this point. The high temperature termination of the Ar Boyle curve is shown, as predicted using the fundamental EOS [4.1] and using the Peng-Robinson EOS, both implemented using the ThermoC software [3.25].

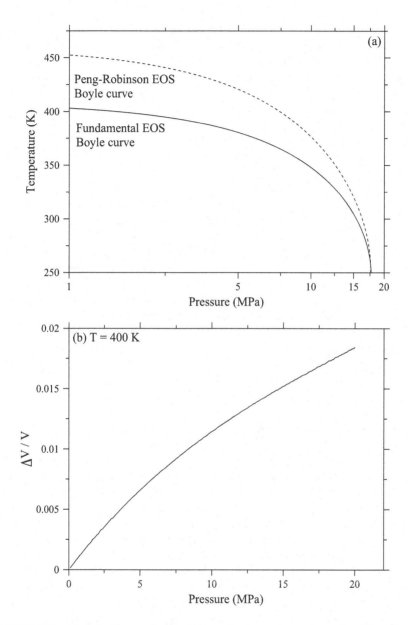

FIGURE 6.13 (a) Ar Boyle curve termination on the temperature axis predicted from the fundamental EOS [4.1] and from the Peng-Robinson EOS, both implemented using the ThermoC software [3.25]. (b) Fractional discrepancy between the volumes predicted by the fundamental EOS and Peng-Robinson EOS for Ar at 400 K.

The fundamental EOS in this region is backed by experimental data and known to be accurate to $\pm$ 0.03% as far as the density is concerned. The discrepancy in the P,V EOS at 400 K between the Peng-Robinson EOS and the fundamental EOS is shown in Figure 6.13 (b); it is less than 2% at 20 MPa, yet the effect on the predicted Boyle curve is drastic.

From examples such as this we can see that, whilst the ability to plot physically realistic inversion curves does not guarantee that an EOS is correct, it is a stringent test of whether an EOS provides a reasonable prediction of fluid P,V,T properties when extrapolated.

6.5 The Frenkel Line

We have seen in previous chapters that fluids confined under P,T conditions close to the melting curve exhibit properties reminiscent of solids. The open question is, therefore, how does the fluid get from P,T conditions where it is a gas to P,T conditions where it is behaving essentially as a solid, given that no first order phase transition is required when $T > T_C$? Is there any narrow change in material properties if a path above T_C is followed or is the variation in properties completely smooth and monotonic throughout? We have seen so far that there is certainly no first order phase transition, and that it is dubious if the Widom lines denote any narrow change in material properties significantly beyond the critical point due to the rapid divergence and smearing out of different Widom lines. In any case, a transition driven by the same underlying physics as the boiling/condensation phase transition cannot, by definition, exist when $T \gg T_C$ as discussed earlier (Section 6.1.4).

An alternate proposal has therefore been put forward in recent years (since 2012 [6.2]). It is supported by tentative experimental evidence obtained by the author [3.24][4.8][4.17] and by other groups [4.13][6.18][6.19]. Based on Frenkel's proposal of the parameter of the liquid relaxation time governing behaviour of the dense fluid (Section 5.1), it is proposed that there is a narrow, though not first order, transition in certain dynamic properties illustrated schematically in Figure 6.14. The transition line is christened the Frenkel line. This line, as shown in Figure 6.14, begins in the subcritical region, but is discussed in this chapter because, in practice, it is often not observable in the subcritical region. It can be obscured by the large maxima in parameters such as c_V that occur in the vicinity of the critical point.

In addition, it is the presence of the Frenkel line in the supercritical region that is most significant. In principle, the Frenkel line continues up to arbitrarily high P,T. In this section we will outline a selection of conditions we can use to define the Frenkel line (Section 6.5.1), compare to the Widom lines (Sections 6.5.2 and 6.5.3) and discuss what, if anything, can lead to the termination of the Frenkel line (Section 6.5.4).

6.5.1 Definitions of the Frenkel Line

We will begin with a fundamental definition of the Frenkel line in terms of the velocity autocorrelation function (VAF). We saw in Section 5.1 how the two different kinds of particle motion in a dense fluid (oscillatory and diffusive) were modelled by Frenkel. The liquid relaxation time τ_R marks the average time that a tagged particle spends in

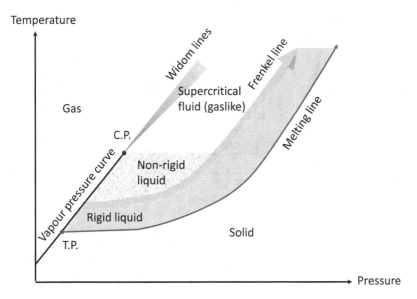

FIGURE 6.14 Schematic fluid phase diagram. The liquid and gas states are separated by the vapour pressure curve, taking in the triple point (T.P.) and terminating at the C.P. The Widom lines extend a finite distance from the C.P. and the liquid state is divided into rigid and non-rigid regions. The latter can persist above T_C on the high density side of the Frenkel line.

an equilibrium position (exhibiting only oscillatory motion) in between jumps to a new equilibrium position (diffusive motion). If we envisage crossing the Frenkel line upon pressure decrease or temperature increase (i.e., crossing over to the gas-like region), the most fundamental definition of the Frenkel line in terms of dynamic properties is that it marks the point at which the oscillatory motion ceases, leaving only the diffusive motion. This condition can be defined mathematically in terms of the VAF as given in equation 6.25 [1.11][1.13]. The velocity autocorrelation function measures (as a function of time) the projection of the velocity v of a tagged particle in the fluid onto its initial value v_0, averaged over the range of initial conditions.

$$Z(t) = \frac{1}{3}\langle v.v_0 \rangle \qquad (6.25)$$

Here, the factor of $1/3$ ensures that the value of the VAF at $t=0$ is simply $k_B T / m$. Diffusive motion of particles is represented by the VAF as a gradual decay whilst oscillatory motion results in oscillations in the VAF. When the Frenkel line is crossed into the gas-like region the minimum in the VAF as a result of the oscillatory motion disappears [1.14][6.20]. The VAF criterion is used to map out the Frenkel line using MD simulations for model fluids [6.20] and for real fluids [6.21]. Figure 6.15 shows MD simulation results for the VAF upon isobaric temperature increase at significantly supercritical conditions for the Lennard-Jones fluid, demonstrating the disappearance of the oscillatory component as the Frenkel line is crossed.

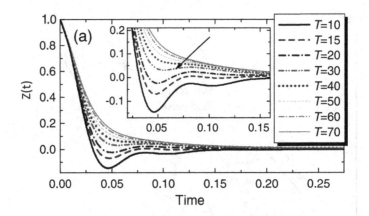

FIGURE 6.15 Calculated VAF ($Z(t)$) for the Lennard-Jones fluid at $P = 200$, demonstrating disappearance of oscillatory component (marked with arrow) upon temperature increase [6.20]. P,T are in Lennard-Jones units.[‡] Reprinted figure with permission from V. V. Brazhkin, Yu. D. Fomin, A. G. Lyapin, V. N. Ryzhov, E. N. Tsiok, and K. Trachenko, Phys. Rev. Lett. **111**, 145901 (2013). Copyright (2013) by the American Physical Society.

In a completely separate investigation from that of the VAF criterion and the heat capacity criterion discussed later that follows from it, the Frenkel line transition has been understood in terms of mesoscopic static structure; it marks the point where particles in a solid-like environment begin to percolate throughout the system [6.22][6.23][6.24].

The most widely used experimental definition of the Frenkel line follows from the VAF criterion. In the high density limit, when the fluid is held under P,T conditions close to the melting curve, a range of shear waves can propagate in the fluid with frequencies reaching from nearly zero up to the Debye frequency (ω_D, with $\tau_D = 2\pi / \omega_D$ as the Debye period), i.e., $\tau_R \gg \tau_D$. This is essentially the same behaviour as that which governs the propagation of shear waves in solids. As we decrease density we expect that the lowest frequency shear waves will no longer be able to propagate in the fluid as τ_R decreases and the particles will not remain stationary for long enough (equation 5.3). Therefore, for shear modes to exist and propagate we require:

$$\tau_R \gtrsim \tau_D \qquad (6.26)$$

As mentioned elsewhere in Section 5.1.2, direct experimental observations of shear waves in dense fluids have been conducted over the years utilizing several different experimental methods. At densities close to crystallization fluids do support high frequency shear waves. However, wide-ranging studies examining in a systematic manner what frequency shear waves can be supported, under what P,T conditions, are lacking.

On the other hand, we implicitly used inequality 6.26 when we used Frenkel's approach to predict the heat capacity of dense fluids in Section 5.1. In equation 5.11 we

---

[‡] Lennard-Jones units are: $T_{LJ} = \frac{k_B T}{\varepsilon}$, and $P_{LJ} = \frac{P\sigma^3}{\varepsilon}$. They are thus the same order of magnitude as, but not identical to, the reduced temperature and pressure P_R, T_R.

considered the contribution to the fluid internal energy arising from propagation of shear modes with frequency ω lying in the range $\omega_F < \omega < \omega_D$. The effect of shear mode propagation on heat capacity allows us to define the Frenkel line also in terms of c_V. The Frenkel line is crossed when no shear wave can propagate (inequality 6.26 is no longer satisfied). We can therefore refer to the higher density side of the Frenkel line as the rigid liquid state. For the simplest case of a monatomic fluid composed of electrically neutral atoms we can define the Frenkel line in terms of heat capacity as the line on the P,T or ρ,T phase diagram along which equation 6.27 is satisfied:

$$c_V = 2k_B \tag{6.27}$$

Equation 6.27 defines the heat capacity that results from the particles having three translational degrees of freedom, with the remaining contribution of $k_B / 2$ to the heat capacity arising from the single longitudinal wave mode. For noble gas fluids the Frenkel line as defined using equation 6.27 can therefore be plotted out over a wide P,T range using experimental data or a fundamental EOS. The heat capacity of other fluids (e.g., liquid metals or molecular fluids) can be studied in an analogous way, provided that other contributions to the heat capacity can be accurately subtracted (see ref. [1.15] and refs. therein). In Appendix B examples are shown where this is the case (for instance N_2) and where it is not the case (for instance CH_4).

In Figure 6.16 we plot the Frenkel line according to the criterion of equation 6.26 for Ne. Unfortunately, the fundamental EOS for Ne is not backed by experimental data to

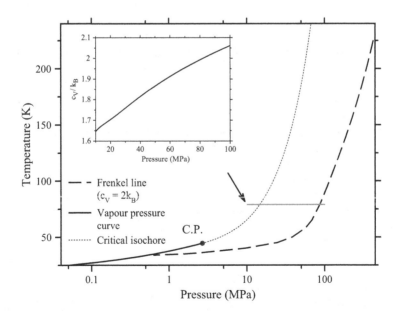

FIGURE 6.16 Vapour pressure curve and Frenkel line for Ne. The Frenkel line is plotted from its intersection with the vapour pressure curve to the highest pressure for which the fundamental EOS is available. Inset: c_V at 75 K. Data are from NIST REFPROP.

high enough pressures at 300 K to observe the condition in equation 6.27. However, a reasonable extrapolation of the $c_V = 2k_B$ line yields a transition pressure in agreement with the X-ray diffraction results shown later (in Figure 6.19). For Ne, it is possible to trace the Frenkel line transition all the way into the subcritical region using equation 6.27, until its eventual termination on the vapour pressure curve. As mentioned earlier, this is not the case for all fluids. Figure 6.17 shows a fluid for which this is not possible: Ar. Panel (a) shows the Frenkel line from the lowest temperature at which it can be reliably calculated using the $c_V = 2k_B$ condition and panel (b) shows the large spike in c_V in the vicinity of the critical point. This spike obscures the continuation of the Frenkel line into the subcritical region for this fluid.

However, the variation in heat capacities along a P,T path crossing the Frenkel line is smooth and monotonic throughout, as shown in the inset to Figure 6.16. There is no fundamental change in the way heat capacity varies as a function of pressure or temperature between gas-like behaviour on one side of the Frenkel line and rigid liquid-like behaviour on the other. This is because, as the Frenkel line is approached from the rigid liquid side the ability to support shear modes is lost gradually as the Frenkel line is approached and τ_R reduces.

Staying with dynamical properties, for molecular fluids an alternate definition of the Frenkel line is in terms of the frequency of the intramolecular vibrational modes which may be studied using Raman or infrared spectroscopy. The pressure dependence of the frequency and linewidth of these modes provides a method to characterize the behaviour of a fluid at both subcritical and supercritical temperature. In a gas, or in a liquid close the critical point, the frequency and linewidth of the spectral peaks generally decrease upon pressure increase. The frequency decrease is due to the increased effect of van der Waals forces between molecules upon density increase (Sections 1.4 and 5.2) and the linewidth decrease is due to collisional narrowing. On the other hand, in a fluid with density close to that of the corresponding solid, the opposite usually happens. The frequency increases upon pressure increase (the same behaviour as that generally observed in solids, Section 5.2) and the linewidth also increases upon pressure increase. The frequency shift can be modelled using the simple Grüneisen model outlined in Section 5.2, whilst there could exist various explanations for the linewidth increase. One explanation proposed by the author is the presence of nonhydrostatic stress in the fluid. In solids, the existence of nonhydrostatic stress always increases linewidth of spectral modes. In a fluid, a static nonhydrostatic stress cannot be supported, but a dynamic nonhydrostatic stress can be supported on any timescale shorter than the liquid relaxation time τ_R. As density is increased, τ_R increases so there is more potential for nonhydrostatic stress to exist in the fluid and lead to broadening of spectral peaks.

The effect of pressure change on the characteristics of spectral peaks is therefore fundamentally different in a gas-like fluid and a dense liquid-like fluid. This is observed in the subcritical region as discussed earlier (Section 5.2), and even far beyond T_C. Figure 6.18 shows the drastic changes in the Raman peak width, integrated intensity, and position trends as the Frenkel line is crossed in CH_4 at 300 K. The transition between these two fundamentally different regimes is always observed at much higher density than the Widom lines and critical isochore. This is seen in the author's studies joining the gas-like and dense liquid-like regimes in a single study [3.24][4.8][4.17], in a small

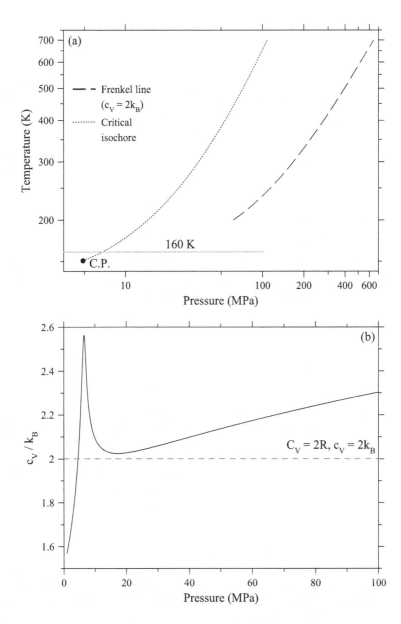

FIGURE 6.17 (a) Frenkel line for Ar (black dashed line) plotted from the lowest temperature at which it can be defined using the $c_V = 2k_B$ criterion up to the highest temperature for which the fundamental EOS is backed by experimental data. The critical isochore is also plotted (black dotted line). (b) Plot of C_V along the isotherm shown at 160 K ($1.06T_C$) showing spike as the critical isochore is approached preventing the $c_V = 2k_B$ criterion from being met. Data are from NIST REFPROP.

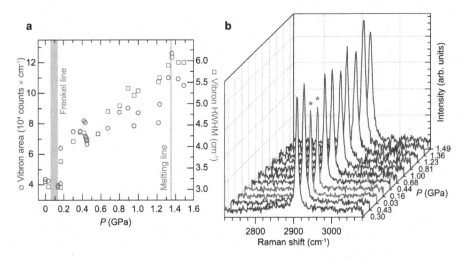

FIGURE 6.18 (a) Width and integrated intensity of CH_4 vibron plotted upon pressure increase across the Frenkel line, and melting curve, at 300 K. (b) Example spectra demonstrating drop in width and integrated intensity on the gas-like side of the Frenkel line (spectra marked with *). Data are from ref. [4.17]. Reprinted figure with permission from D. Smith, M. A. Hakeem, P. Parisiades, H.E. Maynard-Casely, D. Foster, D. Eden, D.J. Bull, A.R.L. Marshall, A.M. Adawi, R. Howie, A. Sapelkin, V.V. Brazhkin, and J.E. Proctor, Phys. Rev. E **96**, 052113 (2017). Copyright (2017) by the American Physical Society.

number of previous studies that significantly predate the proposal of the Frenkel line (e.g., ref. [6.10]) and by joining the findings of previous studies covering the gas-like and liquid-like regimes (as shown earlier in Figure 5.8 for instance).

However, whilst the shifts in Raman frequency and linewidth are easily resolvable on a modern spectrometer, they are small compared to the absolute frequency of the modes (the shifts are typically 0.1 - 0.5%, see Figure 5.8 for instance). In addition, these modes lie at too high a frequency to be excited thermally at accessible temperatures. Therefore, whilst the transition between the regimes (frequency decrease and increase upon pressure increase) is clearly symptomatic of a transition from gas-like to rigid liquid-like behaviour, it does not in itself constitute a direct measurement of a physically relevant property.

The third proposed definition of the Frenkel line relates to static properties rather than dynamic properties. According to conventional wisdom, static properties as measured, for instance, in a diffraction experiment, should all vary smoothly and monotonically when a P,T path significantly above T_C is followed. Recent reliable studies confirm this, with the exception of the co-ordination number (CN). Studies published in recent years, utilizing both X-ray and neutron diffraction, demonstrate changes in CN as a function of pressure which are fundamentally different in the gas-like and rigid liquid-like regimes [4.8][4.13]. Figure 6.19 shows CN data for Ne (collected with X-ray diffraction) and N_2 (collected with neutron diffraction). In both cases, the transition in co-ordination number value lies very close to the Frenkel line as defined by the heat capacity criterion.

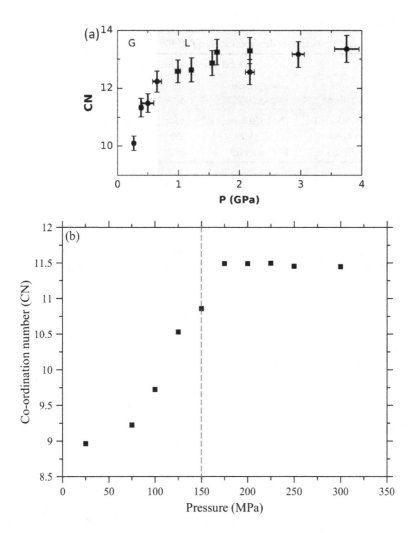

FIGURE 6.19 Variation in co-ordination number (CN) of supercritical fluid Ne measured with X-ray diffraction (a) [4.13] and supercritical N_2 measured with neutron diffraction (b) [4.8] upon isothermal pressure increase at 300 K ($6.7T_C$ for Ne, $2.4T_C$ for N_2). The shaded region in panel (a) corresponds to the liquid-like side of the Frenkel line, and the dashed vertical line in panel (b) corresponds to the Frenkel line position according to the c_V criterion. Panel (a) is reprinted figure with permission from C. Prescher, Yu. D. Fomin, V. B. Prakapenka, J. Stefanski, K. Trachenko, and V. V. Brazhkin, Phys. Rev. B **95**, 134114 (2017). Copyright (2017) by the American Physical Society.

In the gas-like regime the CN steadily increases upon pressure increase. This indicates that, at this stage, it is not necessary for the average distance between a particle and its neighbours to reduce in order to increase the density. In fact, we have seen that for molecular fluids the opposite happens at this stage as far as the intramolecular bond lengths are concerned. To increase density, each particle acquires more nearest

TABLE 6.4   Measurements of Frenkel line position at 300 K made using different criteria outlined in the text. Heat capacity data are from NIST REFPROP unless otherwise stated.

| | Frenkel Line Measurements | | | |
Fluid	VAF (simulated)	Co-ordination Number (CN)	Heat Capacity c_V	Vibrational Mode Frequency
Ne	650 ± 20 MPa [4.13]	750 ± 150 MPa [4.13].	650 ± 20 MPa [4.13]	n/a
N_2	Not measured	165 ± 10 MPa [4.8].	150 MPa	145 ± 20 MPa [4.8].
CH_4	175 MPa [6.21].	Not measured.	n/a	94 ± 80 MPa [4.17].

neighbours. In the dense liquid-like regime however, the density continues to increase whilst the CN stays constant. In this regime, the distance between each particle and its neighbours has to decrease to achieve the necessary density increase.

In principle, the CN should plateau out at a value slightly below that of the sample in the solid state following crystallisation. This was observed in the author's recent study [4.8] but this was maybe just fortuitous—in practice the calculation of the absolute value of the CN is prone to systematic errors due to the choice of where to place the integration limits when the CN is calculated from $g(r)$. It has therefore been proposed [4.38]—and the author agrees—that the relative CN should be considered the sole reliably determined quantity.

In summary, we have discussed the definition of the Frenkel line in terms of one property measurable only through simulations (the VAF) and three properties that are experimentally measurable, that is, heat capacity, vibrational mode frequency (molecular liquids only), and CN. Table 6.4 shows the findings on all fluids in which, to the author's knowledge, the Frenkel line position has been determined with multiple methods in addition to the heat capacity criterion. As we can see, there is reasonable agreement between the findings using the different methods in all cases. However, one could define the Frenkel line in terms of the onset of any solid-like property and study of the extent to which measurements of the Frenkel line position using different properties agree with each other is a subject of ongoing research. Additionally, the Frenkel line has been detected in CO_2 using different neutron diffraction methods in two separate studies [6.18][6.19].

Heat capacity measurements are available from the fundamental EOS over a very wide P,T range and generally agree with simulations of the VAF where available. However, the other measurements of the Frenkel line position in Table 6.4 have been made only at 300 K. Thus, tracking the Frenkel line position as a function of temperature is very much an open field for future research. Unfortunately, the experimental obstacles are formidable; when making optical spectroscopy measurements in the diamond anvil cell (DAC) it is very hard to control and measure pressure with sufficient accuracy above 300 K. A measurement of the Frenkel line position using the CN criterion at a single temperature using X-ray diffraction could easily take an entire day of synchrotron beamtime (subject to the difficulty of controlling and measuring pressure with sufficient accuracy in the DAC), or would take several days of beamtime using apparatus such as a TiZr high pressure cell at a neutron source.

6.5.2 The Frenkel Line and the Widom Lines

We begin our discussion with the location of the Frenkel line and Widom lines at the critical temperature. The Widom lines begin, by definition, at the exact critical point. As we have seen in this chapter, they trend on average to slightly lower density than the critical isochore upon P,T increase. The Frenkel line, by contrast, is crossed at significantly supercritical pressure and density, at the critical temperature. As temperature is increased the pressure at the Frenkel line increases by orders of magnitude in order to ensure that the density at the Frenkel line undergoes a modest increase upon temperature increase, running roughly parallel to the melting curve on the $\log P$, $\log T$ phase diagram. This is the case for all the examples shown in this text, here and in Appendices A and B. The Frenkel line therefore does not come anywhere close to the Widom lines on the P,T or ρ,T phase diagram.

This is no accident; it is due to the physical origin of the Frenkel line as a transition to solid-like (i.e., rigid liquid-like) properties upon density increase. It therefore happens, by definition, at a point where the density is adequate for the fluid to have a relatively close-packed structure. The Widom lines, on the other hand, trend to slightly lower density than the critical isochore. The density at the critical point is such that there is a significant amount of—literally—empty space between adjacent particles in the fluid. As a general rule, there is just enough space to squeeze in an extra particle between two adjacent particles in the fluid at the critical point (it corresponds to the lowest density that a liquid can have, and the sample is also a gas at this point). This is observed experimentally and also predicted theoretically using EOS in which the particle diameter is an input parameter, for instance the Carnahan-Starling, van der Waals EOS, and in simulations using the Lennard-Jones potential [6.25].

We can illustrate this point here using the example of fluid Ne, and refer the reader to other examples in Appendices A and B. Figure 6.20 is a diagram drawn to scale showing

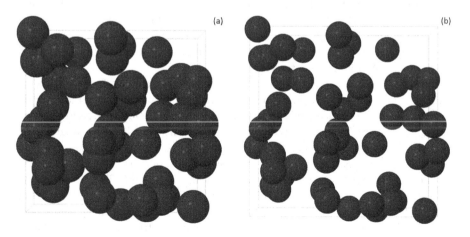

FIGURE 6.20 (a) Fluid Ne at density of 53.744 mol L^{-1}, the density at which the Frenkel line is crossed at T_C (45 K). (b) Fluid Ne at the critical density of 23.882 mol L^{-1}. In both cases Ne atoms are shown at 70% of their full van der Waals radius [6.26] for clarity.

the Ne atoms (with van der Waals radius of 1.54 Å) at the density where the Frenkel line is crossed at T_C (53.744 mol L^{-1}, as defined by the heat capacity criterion $C_V = 2R$, $c_V = 2k_B$ for an atomic fluid), and at the critical density (2.6786 MPa, 23.882 mol L^{-1}). We can see that at the critical point (b) there is a lot of empty space between atoms, compared to when the Frenkel line is crossed (a) (the Ne atoms are shown at 70% of their full van der Waals radius for clarity).

We may also plot out on the P,T phase diagram the critical isochore alongside the Frenkel line in Ne (Figure 6.21 (a)) and compare the Frenkel line and Widom lines positions on the P,ρ EOS at various temperatures (Figure 6.21 (b)). Plotted alongside selected data points from a representative Widom line (that for C_P) the failure of the Frenkel line to stray anywhere near to the critical isochore and Widom lines is illustrated. Whilst the Widom lines stay close to the critical isochore, the Frenkel line exists at significantly higher density than the critical isochore throughout.

6.5.3 Positive Sound Dispersion Above T_C

In Section 5.4.3, we examined the phenomenon of positive sound dispersion (PSD), as a recognised indicator that a sample is in a liquid, or liquid-like, state. This begs the question—is this phenomenon observed above T_C and, if so, under what conditions? PSD has been observed using inelastic X-ray scattering in the DAC under deeply supercritical conditions (for instance, in Ar at 573 K, $T/T_C = 3.8$ [6.3]).

What is now controversial is whether the PSD appears at the Frenkel line or the Widom line. On the one hand, it is tempting to identify the (phenomenological) structural relaxation time delineating the viscous and elastic sound dispersion regimes with Frenkel's liquid relaxation time and thus conclude that PSD will only be observed on the liquid-like side of the Frenkel line. MD simulations conducted prior to the proposal of the Frenkel line have predicted the onset of PSD at conditions coinciding with where we now expect the Frenkel line to lie [6.3], and MD simulations of the Frenkel line position using the VAF criterion found that PSD disappeared close to the Frenkel line [6.20]. To the author's knowledge, all experimental studies identifying PSD in supercritical fluids have observed it only under conditions likely to lie on the liquid-like side of the Frenkel line.

On the other hand, absence of evidence is not evidence of absence. The experimental difficulties involved in a study that would resolve this question one way or the other are formidable. The DAC is designed for experiments at 10 GPa, not 100 MPa. As we decrease pressure across the Frenkel line and towards the Widom line it becomes progressively harder to control, and measure, pressure with sufficient accuracy. An experiment to demonstrate if the onset of PSD occurs at the Widom line or Frenkel line would have to measure and maintain pressure at ca. 50 MPa ± 10 MPa in the DAC. To collect PSD data such as those shown in Figure 5.12, this constant P,T has to held for time in excess of 12 hours [6.27]. In the author's personal experience, it is challenging enough, especially above 300 K, to do this for 5 minutes to collect a Raman spectrum, let alone for 12 hours to collect the IXS data over the range of Q,ω necessary to demonstrate the presence or otherwise of PSD. In addition, molecular dynamics simulations have predicted that PSD should be observed at pressures right down to where the Widom line is expected [6.4].

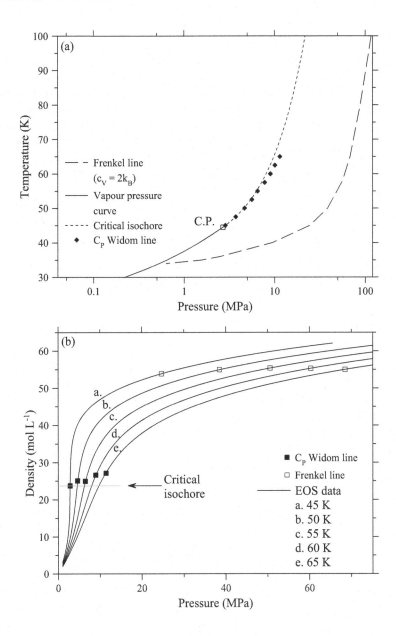

FIGURE 6.21 (a) Widom and Frenkel lines for Ne on the P,T phase diagram. The critical iso-chore of Ne (23.882 mol L^{-1}) is plotted up to 100 K alongside a representative Widom line (that for C_P) plotted to its termination according to the criterion outlined in Section 6.1.1. The Frenkel line for Ne (according to the $c_V = 2k_B$ criterion) is also plotted, from its commencement on the vapour pressure curve up to 100 K. (b) P,ρ EOS of Ne at various temperatures. Densities are marked at which the Widom and Frenkel lines from panel (a) are crossed, as well as the critical isochore. Data are taken from NIST REFPROP.

6.5.4 Termination of the Frenkel Line

The Frenkel line is unique amongst the various transitions that take place in the super-critical region in that there is no physical mechanism for it to terminate. The inversion curves all eventually terminate by intersecting the temperature axis at $P = 0$, the Widom lines get smeared out by $T \sim 2T_C$, but the Frenkel line can in principle continue to arbi-trarily high P,T [6.2]. The only mechanism that the author can see for the Frenkel line to terminate is when the temperature becomes sufficiently high to break the covalent inter-atomic bonds in a molecular fluid and remove the outer electrons from the atoms in any fluid; forming a plasma. Having said that, in some cases the temperatures required for the former are surprisingly experimentally accessible. In Chapter 1 (Figure 1.2) we showed the rise in the heat capacity of hydrogen above 3000 K due to dissociation of the H_2 molecule into atomic hydrogen. This temperature may not be widely relevant to industrial processes, but hydrogen can certainly be subjected to temperatures above this in planetary interiors or in a laser heated DAC experiment.

As previously noted, the experimental studies of the Frenkel line (in both subcritical and supercritical regions) are scarce and an important topic for future research.

6.6 Conclusions

We have seen in this chapter that the supercritical fluid region of the phase diagram, whilst not containing any first order phase transitions, is divided into regions exhibiting gas-like and liquid-like properties by the Frenkel line, extending—so far as we can tell—to arbi-trarily high P,T. The difference between fluids on the liquid-like and gas-like sides of this narrow line is principally in dynamic properties but can also be observed in static proper-ties (co-ordination number). Since the Frenkel line was proposed in 2012 [6.2], its experi-mental confirmation has been controversial [6.28][6.29][6.30]. However, the number of convincing studies demonstrating the existence of the Frenkel line is steadily increasing. In addition, there are studies published prior to the proposal of the Frenkel line that clearly observe this transition (for instance ref. [6.3] and studies cited in ref. [6.31]). Frenkel's con-cept of the liquid relaxation time, on which the Frenkel line is based, has been successfully used to predict the heat capacity of fluids over a wide P,T range (Section 5.1). Thus, at the present time the experimental case for the existence of the Frenkel line is proven, but many further studies are required to elucidate the details of this new transition. As we shall see in Chapter 8, the Frenkel line generally (but not always) lies at densities higher than those rel-evant to the industrial applications of liquids and supercritical fluids. However, we expect it to be of extreme importance to the behaviour of fluids in the interiors of the major and outer planets. This topic is reviewed in Chapter 9.

Even before the proposal of the Frenkel line, the standard textbook view of the supercritical fluid state as a region devoid of any narrow transitions in fluid proper-ties was not really correct. In this chapter we have seen a whole host of Widom lines and inversion curves that extend far into the supercritical region. Some of these tran-sitions are not important, but the Widom lines and Joule-Thomson inversion curve are extremely important from the point of view of industrial applications of fluids. We have seen in this chapter that the Widom lines, known about for decades, can extend

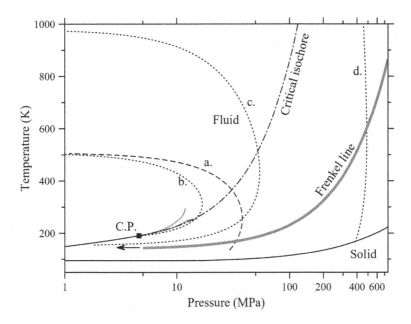

FIGURE 6.22 P,T phase diagram of CH_4. The solid black lines are the vapour pressure curve and melting curve. The dash-dotted line is the critical isochore. Beyond the critical point there exists the Widom lines (representative examples shown with thin grey lines). The thick dashed line is the Zeno curve (a.), the thin dotted lines are the Boyle curve (b.), the Joule-Thomson curve (c.), and the Amagat curve (d.). The thick grey line is the Frenkel line in the range simulated. The arrow indicates the expected continuation of the Frenkel line to meet the vapour pressure curve. Data are from the fundamental EOS via NIST REFPROP (vapour pressure curve, critical isochore and Widom lines), the fundamental EOS via the ThermoC software [3.25] (curves a.–d.), VAF simulations in ref. [6.21] (Frenkel line) and ref. [4.53] (melting curve).

up to ca. $2T_C$ in some cases and the Joule-Thomson inversion curve typically terminates at ca. $5T_C$.

Figure 6.22 summarizes the key message of this chapter with the P,T phase diagram of CH_4. Beyond the critical point lie the Widom lines, Frenkel line, Zeno curve, Boyle curve, Joule-Thomson curve and Amagat curve extending far into the supercritical region.

References

6.1 V.V. Brazhkin and V.N. Ryzhov, J. Chem. Phys. **135**, 084503 (2011).

6.2 V.V. Brazhkin, Yu.D. Fomin, A.G. Lyapin, V.N. Ryzhov and K. Trachenko, Phys. Rev. E **85**, 031203 (2012).

6.3 G.G. Simeoni et al., Nat. Phys. **6**, 503 (2010).

6.4 F.A. Gorelli, T. Byk, M. Krisch, G. Ruocco, M. Santoro and T. Scopigno, Sci. Rep. **3**, 1203 (2013).

6.5 V.V. Brazhkin, Yu.D. Fomin, V.N. Ryzhov, E.E. Tareyeva and E.N. Tsiok, Phys. Rev. E **89**, 042136 (2014).

6.6 V.V. Brazhkin, Yu.D. Fomin, A.G. Lyapin, V.N. Ryzhov, E.N. Tsiok, J. Phys. Chem. B **115**, 14112 (2011).

6.7 M.E. Fisher and B. Widom, J. Chem. Phys. **50**, 3756 (1969).

6.8 R.J.F. Leote de Carvalho, R. Evans, D.C. Hoyle and J.R. Henderson, J. Phys.: Cond. Mat. **6**, 9275 (1994).

6.9 M. Thol, G. Rutkai, A. Köster, R. Lustig, R. Span and J. Vrabec, J. Phys. Chem. Ref. Data **45**, 023101 (2016).

6.10 D. Fabre and R. Couty, C. R. Acad. Sci., Paris, **303**, 1305 (1986).

6.11 F. Reif, *Fundamentals of statistical and thermal physics*, McGraw-Hill, New York (1965).

6.12 U.K. Deiters and A. Neumaier, J. Chem. Eng. Data **61**, 2720 (2016).

6.13 O.L. Boshkova and U.K. Deiters, Int. J. Thermophys. **31**, 227 (2010).

6.14 R. Span and W. Wagner, Int. J. Thermophys. **18**, 1415 (1997).

6.15 V.V. Brazhkin, A.G. Lyapin, V.N. Ryzhov, K. Trachenko, Yu.D. Fomin and E.N. Tsiok, Phys.–Uspekhi **55**, 1061 (2012).

6.16 W. Wagner and A. Pruβ, J. Phys. Chem. Ref. Data **31**, 387 (2002).

6.17 W. Wagner, A. Saul and A. Pruβ, J. Phys. Chem. Ref. Data **23**, 515 (1994).

6.18 C.J. Cockrell et al., Phys. Rev. E **101**, 052109 (2020).

6.19 V. Pipich and D. Schwahn, Phys. Rev. Lett. **120**, 145701 (2018).

6.20 V.V. Brazhkin, Yu.D. Fomin, A.G. Lyapin, V.N. Ryzhov, E.N. Tsiok and K. Trachenko, Phys. Rev. Lett. **111**, 145901 (2013).

6.21 C. Yang, V.V. Brazhkin, M.T. Dove and K. Trachenko, Phys. Rev. E **91**, 012112 (2015).

6.22 T.J. Yoon et al., J. Phys. Chem. Lett. **9**, 6524 (2018).

6.23 T.J. Yoon, M.Y. Ha, W.B. Lee and Y.-W. Lee, J. Phys. Chem. Lett. **9**, 4550 (2018).

6.24 T.J. Yoon, M.Y. Ha, W.B. Lee, Y.-W. Lee and E.A. Lazar, Phys. Rev. E **99**, 052603 (2019).

6.25 D.M. Heyes and L.V. Woodcock, Fluid Phase Equilibria **356**, 301 (2013).

6.26 A. Bondi, J. Phys. Chem. **68**, 441 (1968).

6.27 M. Santoro, private communication.

6.28 V.V. Brazhkin and J.E. Proctor, https://arxiv.org/abs/1608.06883 (2016).

6.29 T. Bryk, F.A. Gorelli, I. Mryglod, G. Ruocco, M. Santoro and T. Scopigno, J. Phys. Chem. Lett. **8**, 4995 (2017).

6.30 V.V. Brazhkin et al., J. Phys. Chem. B **122**, 6124 (2018).

6.31 J. Zhang et al., J. Geochem. Exp. **171**, 20 (2016).

7

Miscibility in the Liquid and Supercritical Fluid States

7.1 Introduction

The miscibility (or lack thereof) of different* liquids and supercritical fluids is a vast subject area. It is not possible to give a good overview of this subject, delivered in a pedagogic manner, in a single chapter; it would require an entire textbook. On the other hand, it is too important a subject to ignore as more or less all of the locations in nature where liquids and supercritical fluids are found, and the applications in industry, involve mixtures of different fluids. Even applications involving nominally pure fluids in practice involve significant amounts of impurity. Theoretically and experimentally there have been major advances in our knowledge of miscibility of different fluids in recent years, some related to the Frenkel line and some not.

Just as we have seen with other properties over the course of this text, the miscibility of liquids and supercritical fluids occupies an intermediate space between that of gases and that of solids. Ideal and real gases are always miscible. On the other hand, solids are usually not miscible. Most possible combinations of different elements in the solid state cannot exist as a single-phase alloy even if adequate energy has been made available (e.g., by quenching from high temperature) for the mixture of atoms to reach its equilibrium state. The miscibility of liquids occupies a middle ground between the cases of solids and gases and, interestingly, has been used as a starting point to understand the miscibility of solids [7.1]. We are all familiar with everyday examples of liquids which are miscible or immiscible. For instance, oil does not mix with water whilst alcohol does.

In this chapter we will therefore look at some basic theory of solution formation to provide background information for a review of recent literature on this topic. Viewing the bibliography is particularly recommended for readers with more than a passing interest in fluid mixtures. We will not cover mixtures of more than two fluids or of fluids that react or decompose; binary mixtures of non-reacting fluids are more than enough

---

* This distinction is important; we do not consider here mixtures of different states of the same substance such as a mixture of liquid water and water vapour.

for one chapter. Having said that, some of the results presented here can be readily generalized to mixtures of an arbitrary number of non-reacting fluids.

Experimentally, one can see at a glance that the parameter space to be explored is vast. Even for a binary mixture of two fluids which do not react with the other fluid or decompose in its presence one must consider, for a sample with a fixed composition, the behaviour of the fluid mixture, coexisting with regions in which one of the other of the component fluids exist in phase-separated form. All three of these regions can undergo separate transitions between different states. So, with the exception of a limited set of industrially important fluid mixtures under modest pressures (section 7.4), humankind has not explored much of this parameter space, and these experiments are fraught with difficulties. In Section 7.5 we will discuss some experimental methods used to study fluid mixtures.

7.2 Raoult's Law, Henry's Law, and the Lever Rule

7.2.1 Raoult's Law and Henry's Law

We will begin with the limiting cases in which the miscibility of two fluids is effectively modelled by two laws: Raoult's law and Henry's law, covering these by loosely following the approach taken in ref. [7.2]. Raoult's law is derived on the assumption that the intrinsic rate of evaporation of both components in the binary fluid is independent of fluid composition. So, if we have a 50-50 at.% fluid mixture of components 1 and 2, then the rate of evaporation of the component 1 particles is half the rate of evaporation from a pure sample of component 1 with the same surface area at the same P,T conditions. We may specify the conditions that must be met by the function giving the potential energy $V_{ij}(r)$ between particles of components i and j at separations close to the equilibrium separation r_{eq} (equation 7.1).

$$V_{11}(r_{eq}) \approx V_{12}(r_{eq}) \approx V_{22}(r_{eq}) \tag{7.1}$$

In this case, the similar depth of the different potential wells ensures that the intrinsic rate of evaporation remains constant. We thus begin by considering the vapour-liquid equilibrium for the separate components 1 and 2 of the mixture. We will denote the mole fractions of liquid i in the solution by X_i The rate of evaporation of component i (when $X_i = 1$) is denoted by e_i and is determined by the cohesion of the liquid, i.e., depth of V_{ii} potential well etc. The rate of condensation of component i, c_i, is determined by the rate at which particles from the vapour strike the surface of the liquid, giving equation 7.2 where k_i is a constant and P_i^0 is the vapour pressure of the pure fluid i. At equilibrium we therefore write:

$$e_i = c_i = k_i P_i^0 \tag{7.2}$$

Now consider adding a small proportion of liquid 2 to liquid 1. We will denote the mole fractions of liquid i in the solution by X_i. The rate of evaporation of liquid 1 from the solution will be denoted by e_1^S and the partial vapour pressure of component i by P_i. At equilibrium we may write:

$$e_1 X_1 = e_1^S = k_1 P_1 \tag{7.3}$$

Combining equations 7.2 and 7.3 we obtain Raoult's law, equation 7.4.

$$P_1 = X_1 P_1^0 \tag{7.4}$$

Let us now derive Henry's law. Henry's law applies to liquids which are less miscible. In terms of the potentials $V_{ij}(r)$ we can describe this condition as follows. Close to the equilibrium separation between particles r_{eq}, the cohesion of a particle from liquid 1 surrounded by particles from liquid 2, or vice versa, is weak compared to the cohesion of particles in the pure liquids 1 and 2. So in the former case the liquid particle is sitting in a much shallower potential well[†]. We write:

$$\left| V_{12}\left(r_{eq}\right) \right| \ll \left| V_{ii}\left(r_{eq}\right) \right| \tag{7.5}$$

We represent this situation mathematically as if the intrinsic rate of evaporation of component 1 has increased from e_1 to e_1'. Thus equation 7.3 is modified to:

$$e_1' X_1 = e_1^S = k_1 P_1$$

Substituting for k_1 with reference to the actual intrinsic evaporation rate ($e_i = k_i P_i^0$) we obtain an alternate relation to Raoult's law, equation 7.6. This is Henry's law. Here, k_1' is a constant for a given solute and tabulated data are available in many cases [7.3].

$$X_1 k_1' = P_1 \tag{7.6}$$

Where $k_1' = \frac{e_1'}{e_1} P_1^0$.

For the case of poorly miscible liquids we expect $e_1' / e_1 > 1$ so, for a given value of X_1 Raoult's law gives a higher partial vapour pressure P_1. It is common for weakly miscible liquids to exhibit a crossover between Henrian and Raoultian behaviour depending on X_1. For small X_1 the liquid 1 particles in the mixture are sitting in a potential well dominated by the form of $V_{12}(r)$ obeying inequality equation 7.5, resulting in Henrian behaviour. As X_1 increases, more of the liquid 1 particles have nearest neighbours also from liquid 1 and sit in a potential dominated by the form of $V_{11}(r)$. Hence liquid 1 exhibits Raoultian behaviour in this case. The crossover between Henrian and Raoultian behaviour is shown schematically in Figure 7.1. We have not assumed that there is anything "special" about liquid 1 compared to liquid 2, so the same arguments can be made regarding the behaviour of liquid 2 as a function of its mole fraction X_2. Since, by definition $X_2 = 1 - X_1$, we can observe that when liquid 1 is obeying Raoult's law liquid 2 obeys Henry's law and vice versa.

Throughout, we have assumed that we can define vapour pressures and are thus in the subcritical regime. The theory of solubility of supercritical fluids at liquid-like densities is discussed later but is still at an early stage of development.

[†] Though it is less relevant in the present context, we can also apply Henry's law to fluids where the inequality in equation 7.5 is the other way round: $\left| V_{12}\left(r_{eq}\right) \right| \gg \left| V_{ii}\left(r_{eq}\right) \right|$

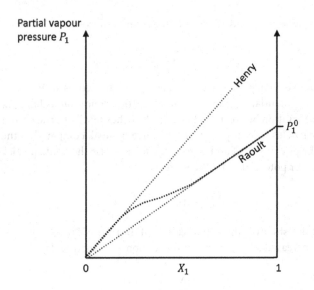

FIGURE 7.1 Schematic diagram showing crossover of solute in a binary solution from Raoultian to Henrian behaviour as a function of molar fraction X_1.

7.2.2 Change in Gibbs Function on Mixing of Raoultian Liquids

The appropriate thermodynamic potential to consider in relation to solution formation is the Gibbs function; two fluids will mix if it leads to a reduction in the value of the total Gibbs function for the system. We can see that this is always the case when Raoult's law is being obeyed. To obtain the change in the Gibbs function upon mixing we consider the addition of fluid 1 to fluid 2 as a three-stage isothermal process [7.2]:

1. Evaporation of the fluid 1 at vapour pressure P_1^0 (if we are in the subcritical regime).
2. A decrease in pressure (with no phase change) to the pressure at which fluid 1 will enter solution, P_1.
3. The condensation of fluid 1 into solution at this pressure, which becomes the partial pressure of fluid 1 in the solution.

By definition, steps 1 and 3 take place when the Gibbs function for the liquid and gas phases are equal so involve no change in the Gibbs function. Thus the change in the Gibbs function is due solely to stage 2. Providing this takes place at P, T conditions where the ideal gas EOS may be employed this leads to a simple expression for the change in the Gibbs function upon solution formation. Using $G = U - TS + PV$ and $dU = TdS - PdV$ we obtain $dG = VdP$ and hence, for 1 mole:

$$\Delta G = \int_{P_1^0}^{P_1} VdP = \int_{P_1^0}^{P_1} \frac{RT}{P} dP = RT \ln\left(\frac{P_1}{P_1^0}\right)$$

Considering both fluids in the binary solution, the total change in the Gibbs function ΔG_{MIX} due to solution formation is:

$$\Delta G_{MIX} = RT\left[X_1 \ln\left(\frac{P_1}{P_1^0} \right) + X_2 \ln\left(\frac{P_2}{P_2^0} \right) \right] \tag{7.7}$$

For a binary solution obeying Raoult's law we can equate the partial pressures to the mole fractions (equation 7.4) after which, substituting for X_2 gives us:

$$\Delta G_{MIX} = RT\left[X_1 \ln X_1 + (1 - X_1)\ln(1 - X_1) \right] \tag{7.8}$$

Thus, in this case components 1 and 2 are miscible in all ratios since $\Delta G_{MIX} < 0$ by definition according to equation 7.8 (plotted in Figure 7.2). Note that in this case we have assumed that both components of the solution are obeying Raoult's law (as opposed to the case earlier when one obeys Raoult's law and one obeys Henry's law). In addition, mixing can still happen if it is endothermic (due to the increase in entropy on mixing) since from the above definitions we can obtain equation 7.9 relating ΔG_{MIX} to the change in internal energy on mixing, ΔU_{MIX}.

$$\Delta U_{MIX} = T\Delta S_{MIX} + \Delta G_{MIX} \tag{7.9}$$

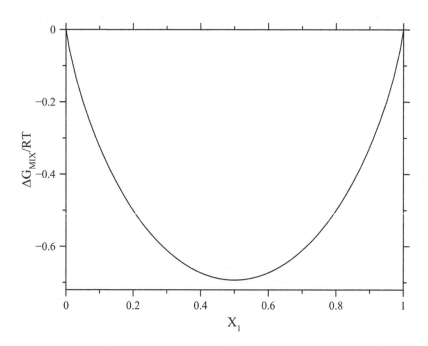

FIGURE 7.2 Change in Gibbs function on formation of a binary solution where both components obey Raoult's law, as a function of the mole fraction of one component in the solution.

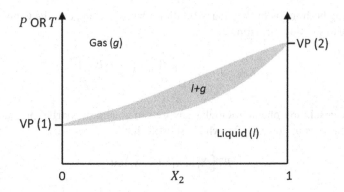

FIGURE 7.3 Phase diagram of two miscible fluids as X_2 and either P or T is varied in the sub-critical domain for both fluids. As shown, the y-axis represents either increasing T or decreasing P. At $X_2 = 0,1$ the vapour pressure curve of the respective pure fluids is crossed (VP (1,2)) whilst for $0 < X_2 < 1$ there is a region of coexistence of liquid and gas states (l+g) with fluid proportions governed by the lever rule.

7.2.3 Phase Equilibria in Miscible Fluids: The Lever Rule

The phase equilibria in fluids which are completely miscible can be described using simple analytic expressions; the lever rule for the subcritical case. Figure 7.3 illustrates schematically the phase diagram of such a mixture in which we vary either pressure or temperature, in a range which crosses the vapour pressure curve of both fluid components in the subcritical regime.

The lever rule[‡], applying in the mixed phase (l+g) region, is derived by considering the mole fractions of the fluid mixture that are in the liquid (n_l) and gas (n_g) states in the mixed phase region. This gives the total mole fraction X_1 of fluid 1 in the system, in terms of the mole fractions of fluid 1 that are in the liquid and gas states (X_{1g}, X_{1l}):

$$X_1 = n_g X_{1g} + n_l X_{1l}$$

For the case where the P, T path followed crosses into the supercritical region of one of the mixture components things are slightly more complex, but some properties can still be treated analytically [3.4].

7.3 Hildebrand Theory of Mixing

7.3.1 Internal Energy of Fluid Mixtures Using Hildebrand Theory

The Hildebrand theory (outlined in ref. [7.4]) allows us to predict the change in Gibbs function on mixing for non-Raoultian solutions but still with some

---

[‡] It is referred to as the lever rule because it can also be written $n_g(X_1 - X_{1g}) = n_l(X_{1l} - X_1)$ in which it bears a mathematical resemblance to the lever rule from mechanics.

assumptions (explored below) about the nature of the solution. A solution following these assumptions is referred to as a regular solution. We begin with the expression below for the part of the pure fluid internal energy per mole that is due to the interaction between particles U_{int} (extracted from the energy equation[§], equation 4.15), modify it for a fluid mixture, and from this derive the change in internal energy on mixing. This is then combined with the assumption of ideal entropy change on mixing (explored later) to obtain an expression for the change in the Gibbs function upon mixing.

$$U_{int} = 2\pi N_A \rho \int_{r=0}^{\infty} V(r)g(r)r^2 dr$$

Here, to obtain the internal energy per mole we are multiplying by Avogadro's number. To consider a binary fluid mixture we need to consider three separate integrals arising from the interactions of like and unlike fluid particles and assume that the total fluid volume is obtained by summing the volumes occupied by the mole fractions of fluids 1 and 2 (though readers may recall being shown fluid mixtures in high school for which this assumption is not quite valid!). Assuming $g_{12}(r) = g_{21}(r)$ and $V_{12}(r) = V_{21}(r)$ we obtain:

$$U_{int} = \frac{2\pi N_A}{X_1 V_M(1) + X_2 V_M(2)} \left[\begin{array}{l} X_1^2 \int_0^\infty V_{11}(r)g_{11}(r)r^2 dr + 2X_1 X_2 \int_0^\infty V_{12}(r)g_{12}(r)r^2 dr \\ + X_2^2 \int_0^\infty V_{22}(r)g_{22}(r)r^2 dr \end{array} \right] \tag{7.10}$$

To proceed it is necessary to assume that every $V_{ij}(r)$ function has the mathematical form of the Lennard-Jones potential (equation 2.4). In this case all $V_{ij}(r)$ functions can be written in terms of the dimensionless separation $r_D = r/a_{ij}$. As a result, the $g_{ij}(r)$ functions become identical when written in terms of r_D.

$$V_{ij}(r_D) = 4\varepsilon_{ii} \left[\frac{1}{r_D^{12}} - \frac{1}{r_D^6} \right] = 4\varepsilon_{ij} V_D(r_D)$$

$$g_{ij}(r) = g(r_D) \text{ for all } i \tag{7.11}$$

In this case, the integrals in equation 7.10 become identical and we obtain:

$$U_{int} = \frac{8\pi N_A}{X_1 V_M(1) + X_2 V_M(2)} \left[X_1^2 \varepsilon_{11} a_{11}^3 + 2X_1 X_2 \varepsilon_{12} a_{12}^3 + X_2^2 \varepsilon_{22} a_{22}^3 \right] \int_0^\infty V_D(r_D)g(r_D)r_D^3 dr_D$$

[§] Recall that here ρ refers to the number density rather than the mass density.

We now calculate the change in internal energy upon mixing, denoting the integral in the equation above by I and the volume fractions of the component fluids by $V_F(1), V_F(2)$:

$$\Delta U_{MIX} = U_{int} - X_1 U_{int}(X_1 = 1, X_2 = 0) - X_2 U_{int}(X_1 = 0, X_2 = 1)$$

$$\Delta U_{MIX} = 8\pi N_A I \left[X_1 V_M(1) + X_2 V_M(2) \right] \left[\frac{2\varepsilon_{12}a_{12}^3}{V_M(1)V_M(2)} - \frac{\varepsilon_{11}a_{11}^3}{V_M(1)^2} - \frac{\varepsilon_2 a_{22}^3}{V_M(2)^2} \right] V_F(1) V_F(2)$$

It is most useful to express this in terms of the energy of vaporization, and hence cohesive energy density, of the component fluids. Hildebrand proposed that $\Delta U_{VAP}(i) = -U_{int}(X_1 = 1, X_2 = 0)$. By definition, the cohesive energy density of fluid i ($C^{(i)}$) is the energy of vaporization per unit volume, resulting in:

$$\Delta U_{MIX} = V_F(1) V_F(2) \left[X_1 V_M(1) + X_2 V_M(2) \right] \left[\sqrt{C^{(1)}} - \sqrt{C^{(2)}} \right]^2$$

$$\text{Where } C^{(i)} = \Delta U_{VAP}(i) / V_M(i) \tag{7.12}$$

Equation 7.12 is the principal result of Hildebrand's theory. What does it tell us? With the approximations made above, $\Delta U_{MIX} > 0$ in all cases, i.e., mixing is endothermic. However, this energy cost can be minimized when the component fluids have similar cohesive energy density so $\left[\sqrt{C^{(1)}} - \sqrt{C^{(2)}} \right]$ is minimized. In this case the increase in entropy upon mixing ensures that $\Delta G_{MIX} < 0$ despite $\Delta U_{MIX} > 0$ (equation 7.9). We can intuitively understand why $\left[\sqrt{C^{(1)}} - \sqrt{C^{(2)}} \right]$ must be minimized by considering the other limiting case, where $C^{(1)} \gg C^{(2)}$. In this case the sole effect of adding the fluid 2 particles would be to reduce the cohesion of the fluid 1 particles by pushing them apart, at significant cost in energy. Since the term $\sqrt{C^{(i)}}$ (rather than $C^{(i)}$) occurs in equation 7.12, this quantity has its own name, the solubility parameter.

For the case of an ideal solution, the entropy increase due to mixing is given by a simple analytical expression $\left(\Delta S_{MIX} = -R(X_1 \ln X_1 + X_2 \ln X_2) \right)$ [7.2]. We can combine this result with equation 7.12 to predict whether different fluids are miscible. Methods are also available to deal with the residual entropy [7.5]. We simply need the cohesive energy densities for the pure fluids (there are tables of these [7.6]) and faith in the assumptions. We will return to the topic of cohesive energy density later when we consider the Raman spectra of fluid mixtures.

7.3.2 *P, V, T* EOS for Mixtures Using Hildebrand Theory

Elements of the Hildebrand mixing theory are utilized to obtain simple P,V,T EOS for mixtures. Following the approach outlined in ref. [3.4] we can illustrate this approach beginning with the pressure equation (equation 4.15), reproduced below:

$$\frac{P}{\rho k_B T} = 1 - \frac{2\pi\rho}{3k_B T} \int\limits_{r=0}^{\infty} \frac{dV}{dr} g(r) r^3 dr$$

For a binary mixture we can divide the integral into terms depending on the different $V_{ij}(r), g_{ij}(r)$ analogously to equation 7.10:

$$\frac{P}{\rho k_B T} = 1 - \frac{2\pi\rho}{3k_B T} \left[\begin{array}{c} X_1^2 \int\limits_{r=0}^{\infty} \frac{dV_{11}}{dr} g_{11}(r) r^3 dr + 2X_1 X_2 \int\limits_{r=0}^{\infty} \frac{dV_{12}}{dr} g_{12}(r) r^3 dr \\ + X_2^2 \int\limits_{r=0}^{\infty} \frac{dV_{22}}{dr} g_{22}(r) r^3 dr \end{array} \right] \tag{7.13}$$

Before we apply Hildebrand mixing theory to this we should mention, for completeness, two even more crude approximations that may suffice under certain limited conditions. At low density, if we use the virial EOS (equation 1.2) for the mixture and retain terms up to $B_2(T)$ only, the integral from the pressure equation above determines the exact value of $B_2(T)$:

$$B_2(T) = \frac{2\pi N_A}{3k_B} \left[\begin{array}{c} X_1^2 \int\limits_{r=0}^{\infty} \frac{dV_{11}}{dr} g_{11}(r) r^3 dr + 2X_1 X_2 \int\limits_{r=0}^{\infty} \frac{dV_{12}}{dr} g_{12}(r) r^3 dr \\ + X_2^2 \int\limits_{r=0}^{\infty} \frac{dV_{22}}{dr} g_{22}(r) r^3 dr \end{array} \right]$$

Alternately, we may assume at any density that the fluids composing the mixture consist of very similar particles such that $V_{11}(r) = V_{12}(r) = V_{22}(r)$ and likewise for $g_{ij}(r)$. This is referred to as the "random mixing" approximation and results in a single pressure equation for the mixture [3.4].

A more rigorous approach than these two approximations is to apply Hildebrand's methods outlined above to equation 7.13, specifically the rescaling of $V_{ij}(r)$ and $g_{ij}(r)$ (equation 7.11). This results in:

$$\frac{P}{\rho k_B T} = 1 - \frac{8\pi\rho}{3k_B T} \left[\varepsilon_{11} a_{11}^3 X_1^2 + 2\varepsilon_{12} a_{12}^3 X_1 X_2 + \varepsilon_{22} a_{22}^3 X_2^2 \right] \int\limits_{r=0}^{\infty} \frac{dV_D}{dr_D} g(r_D) r_D^3 dr_D \tag{7.14}$$

In equation 7.14 the integral has a value independent of P, T, X_i so we have reformulated the problem into one of predicting the various ε_{ij}, a_{ij}. The parameters ε_{ii}, a_{ii} are properties of the pure fluids so there is usually a range of thermodynamic and scattering data in the literature from which these parameters can be obtained. However, given the very wide range of possible mixtures that is unlikely to be the case for ε_{ij}, a_{ij} where $i \neq j$. So how can these be estimated? A multitude of empirical and semi-empirical mixing rules have been developed over the decades, some of which are listed in Table 7.1 and reviewed elsewhere [3.4][7.7][7.8].

For many applications these rules do not give adequate accuracy. If this is the case, then what are the alternatives? Experimental measurement of ε_{ij}, a_{ij} for a specific fluid mixture is likely to involve more effort than can be justified unless the problem is one of extreme industrial importance but atomistic simulation to obtain ε_{ij}, a_{ij} is an alternative approach that becomes more feasible as computers become more powerful [7.5][7.7][7.9].

TABLE 7.1   A selection of mixing rules to predict parameters ε_{12}, a_{12}.

Rule	ε_{12}	a_{12}
Berthelot	$\varepsilon_{12} = \sqrt{\varepsilon_{11}\varepsilon_{22}}$	$a_{12} = \sqrt{a_{11}a_{22}}$
Lorentz	$\varepsilon_{12} = \dfrac{1}{2}(\varepsilon_{11} + \varepsilon_{12})$	$a_{12} = \dfrac{1}{2}(a_{11} + a_{12})$
Berthelot-Lorentz ($\delta_\varepsilon, \delta_a$ are small)	$\varepsilon_{12} = \delta_\varepsilon \sqrt{\varepsilon_{11}\varepsilon_{22}}$	$a_{12}^3 = \dfrac{1-\delta_a}{2}\left(a_{11}^3 + a_{22}^3\right)$
Waldman and Hagler	$\varepsilon_{12} = 2\sqrt{\varepsilon_{11}\varepsilon_{22}} \dfrac{a_{11}^3 a_{22}^3}{a_{11}^6 + a_{22}^6}$	$a_{12} = \left(\dfrac{a_{11}^6 + a_{22}^6}{2}\right)^{\frac{1}{6}}$

7.4 Application of the Fundamental EOS to Mixtures

The fundamental EOS method discussed at length in Section 3.3 has also been applied to fluid mixtures. This research has focussed primarily on natural gas mixtures, for instance in the development of the GERG-2008 EOS for mixtures of 21 different natural gas components [3.18]. EOS have also been developed for mixtures relevant to carbon capture and storage technology [7.10].

The data used to produce the fundamental EOS for the mixture are the fundamental EOS for each fluid component, the critical parameters of each component, and the available data on binary mixtures of the different components. The resulting fundamental EOS predicts the Helmholtz function for mixtures of all possible proportions of the different components across a certain temperature and pressure range. However, it is fitted to experimental data for only a tiny proportion of this vast parameter space. Generally, these EOS are only claimed to be valid up to modest pressure (for instance 35 MPa for the GERG-2008 EOS). Therefore, they can only operate at liquid-like densities achieved through low temperature, not high pressure.

We will outline here the methodology used in the generalization of the pure fluid fundamental EOS to produce EOS for mixtures such as the GERG-2008, using notation consistent with that employed elsewhere in this text. We will construct a fundamental equation for a mixture of up to N components, where X_i denote the mole fractions of each component in the mixture. We will use the notation X to denote the mole fractions of all components and ρ_C, T_C to denote the pure fluid critical parameters for all separate components in the mixture.

The ideal gas part of the Helmholtz function for the mixture is relatively simple to calculate by summing the (non-reduced) ideal gas parts of the fundamental EOS for each component $F_I^{(i)}(\rho, T)$ (defined earlier in equation 3.9) and adding the contribution due to the entropy increase on mixing [7.2]. We obtain, for the dimensionless ideal gas component of the mixture Helmholtz function:

$$F_I^{\prime(mix)} = \sum_{i=1}^{N} X_i \left[\frac{F_I^{(i)}(\rho, T)}{RT} + \ln X_i \right] \tag{7.15}$$

The input parameters in this equation are the absolute values of ρ, T for the mixture; there has been no reduction or rescaling of these parameters. On the other hand, rescaling of ρ, T into parameters measured relative to their values at the critical point was an integral part of the construction of the residual Helmholtz function for the single-component fluid (equation 3.12); it allowed the inclusion of Gaussian lineshape components to reproduce the drastic variation in properties close to the critical point.

For the mixture, residual reduced Helmholtz functions for each component are summed in proportion to the different mole fractions X_i. However these are calculated using ρ, T relative to reducing functions $\rho_R(X, \rho_C, T_C), T_R(X, \rho_C, T_C)$ calculated from the mole fractions and pure fluid critical parameters, instead of rescaling relative to any real critical parameters. Analogously to equation 3.9 we write:

$$\rho'^{(mix)} = \frac{\rho}{\rho_R(X, \rho_C, T_C)}$$

$$\mu^{(mix)} = \frac{T_R(X, \rho_C, T_C)}{T}$$

The parameters $\rho_R(X, \rho_C, T_C), T_R(X, \rho_C, T_C)$ are obtained using what are essentially mixing rules, but are far more complex than those employed in the Hildebrand approach to EOS for mixtures (Table 7.1). The residual reduced Helmholtz function for the mixture is thus:

$$F_R'^{(mix)} = \sum_{i=1}^{N} X_i F_R'^{(i)}\left(\rho'^{(mix)}, \mu^{(mix)}\right) + \sum_{i=1}^{N-1} \sum_{j=i+1}^{N} X_i X_j F_{ij} \alpha_{ij}^r\left(\rho'^{(mix)}, \mu^{(mix)}\right)$$

The double summation is known as the "departure function," consisting of F_{ij} which are constants for each possible binary mixture and $\alpha_{ij}^r\left(\rho'^{(mix)}, \mu^{(mix)}\right)$ which is a complex empirical function of $\rho'^{(mix)}, \mu^{(mix)}$, but not of X.

As presented here (for instance in the entropy of mixing term in equation 7.15) it is assumed that all components are perfectly mixed. But what if the mixture is unstable against a phase separation(s)? This extremely complex topic is the subject of ongoing research.

7.5 Some Comments on Experimental Study of Supercritical Fluid Mixtures

7.5.1 Preparation of Fluid Mixtures in the Diamond Anvil Cell (DAC)

At pressures not too far from ambient conditions a variety of methods have been developed to study fluid mixtures (see reviews in refs. [1.5][3.4]). However, at GPa pressures in the diamond anvil cell there are specific experimental difficulties that deserve to be mentioned. Due to the compressibility of gases and the extreme pressures attained in the DAC, any sample which is loaded as a gas at ambient conditions will compress to an infinitesimally small sample size at GPa pressures. In Appendix D we discuss the methods developed to overcome this problem, methods which are fraught with (even more) difficulty when applied to fluid mixtures (Section D.2.2).

To compound these difficulties, due to the minute sample size in the DAC (~ 200 μm diameter sample chamber) it is not possible to make accurate measurements of mixture composition in situ, pre- or post-experiment. The size of the error bars on Figure 7.4 later are a good illustration of the challenges involved in performing DAC experiments on fluid mixtures.

7.5.2 Raman Spectra of Fluid Mixtures; Cohesive Energy Density

Raman spectroscopy can be a valuable experimental method to study mixtures in which at least one component is a molecular fluid. In this case the Raman peaks of the fluids are typically shifted in energy relative to their position in the pure fluid under the same P,T conditions. Predicting these shifts quantitatively is hard but a reasonable understanding can be gained from looking at the cohesive energy density (CED) of the pure fluids and the fluid mixture. The CED of the mixture is estimated (equation 7.16 below) from the volume fractions $V_F^{(i)}$ of the component fluids rather than the molar fractions. We will denote the CEDs using $C^{(i)}$ and C^{MIX}. Table 7.2 lists the CEDs of a variety of different fluids; we can see that mixtures with a wide range of CEDs can be created from different fluids.

$$C^{MIX} = \frac{X_1 V_M(1)C^{(1)} + X_2 V_M(2)C^{(2)}}{X_1 V_M(1) + X_2 V_M(2)} \tag{7.16}$$

CED has units of pressure. It has been proposed that, as far as the effect on individual molecules is concerned, it has an effect roughly equivalent to a hydrostatic pressure. The evidence supporting this assertion is that, in many cases (but not all), the shift in Raman peak positions from molecules caused by a change in CED when the molecules are in solution with another fluid is similar in magnitude to the shift caused by a change in hydrostatic pressure [7.11][7.12]. This effect can be modelled effectively using MD

TABLE 7.2 Cohesive energy density (CED) of a range of liquids (tabulated data from ref. [7.6]).

Liquid	CED (MPa) [7.6]
Argon	Negligible
Hexane	222.0
Acetone	412.1
Ethanol	676.0
Glycerol	1142.4
Water	2294.4

simulations [7.11][7.13]. One can even create a negative pressure by mixing a fluid with another fluid that has a lower CED, resulting in a lengthening of the intra-molecular bonds and downward shift in Raman peak position [7.14]. On the other hand, H_2O has an unusually high CED of 2294 MPa, but you do not feel a pressure of 2294 MPa when you dip your finger in a glass of water.

So, to what extent is CED a real pressure? What happens if we take, for instance, a sample of H_2O (CED 2294 MPa) and add in one molecule of $(CH_3)_2CO$ (acetone, CED 412 MPa)? In this case we have pushed apart the H_2O molecules which are strongly bound together to squeeze in a $(CH_3)_2CO$ molecule, which will then be squeezed by the surrounding water molecules and have reduced bond lengths.

Probably, CED does act as a real pressure on some objects larger than single molecules. For instance, the pressure-induced gelation of individual potato starch grains (diameter ~ 10 μm) occurs at different pressures in different liquid mixtures, correlating well with the CED of the mixtures [7.12]. Some initial experiments suggested that the same may be true for the Raman peak positions of bundled single-walled carbon nanotubes [7.15], though later experiments by this author showed that this correlation did not hold for a wide range of fluids [7.16].

On first inspection, mixing fluids with significantly different CEDs appears to be a method through which some extremely interesting experiments can be done, where the molecules of the fluid component with a higher CED are subjected to negative pressure. Unfortunately, as outlined earlier in the derivation leading to equation 7.12, a large difference in CED between fluid components leads to the fluids being immiscible.

7.6 Open Questions in the Study of Dense Fluid Mixtures

In this final section, we discuss some open questions in the study of dense fluid mixtures that may interest readers. Further discussion on current research on dense fluid mixtures is presented in Chapter 9, with reference to the major and outer planets.

7.6.1 Is Hydrophobicity an Absolute Property?

Most of us are familiar with the fact that water and oil do not mix from time spent in the kitchen. The theory outlined above correctly predicts this lack of miscibility. Water has an extremely high CED whilst hydrocarbons without -OH groups attached exhibit relatively low CED (the example of hexane given in Table 7.2 is representative). Thus equation 7.12 predicts that mixing would be strongly endothermic. We therefore describe hydrocarbons as hydrophobic.

On the other hand, it has been shown recently [7.17] that under GPa pressure the simplest hydrocarbon, CH_4, is no longer hydrophobic (Figure 7.4). The solubility of CH_4 in H_2O is shown to drastically increase beyond that dictated by Henry's law above ca. 1.5 GPa up to a plateau at ca. 0.45 mol%. Clearly at some point we would expect to see a crossover to a regime where, instead of the CH_4 component obeying Henry's law the H_2O component is obeying Henry's law (i.e., the H_2O is weakly soluble in CH_4 rather

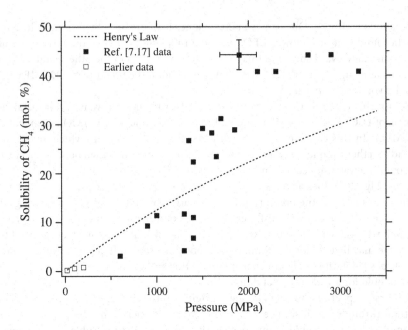

FIGURE 7.4 Solubility (in mol. %) of CH_4 in H_2O at 373 K (ref. [7.17] and refs. therein). Filled squares are the experimental datapoints from ref. [7.17], open squares are earlier data and the dotted line is the solubility expected from Henry's law. Error bars are given on only one datapoint for clarity.

than the other way round, cf. Section 7.2). However, this crossover would be observed at much higher CH_4 concentrations so is not what is observed in Figure 7.4.

What causes the observed change? Arguments have been put forward regarding the reduction in the absolute size difference between the CH_4 and H_2O molecules upon pressure increase but these do not really address the key reason why hydrocarbons are hydrophobic—the presence/lack of hydrogen bonding and the large difference in CED that results. Classical MD simulations have been attempted to explain this effect but the simulations at present do not explain the observed effect [7.18]. So this whole matter is wide open for further experimental and theoretical investigation. See also C.G. Pruteanu et al., J. Phys. Chem. Lett. **11**, 4826 (2020).

7.6.2 Miscibility in the Supercritical Fluid State

Ideal and real gases are always miscible. We have spent most of this chapter discussing the limited miscibility of liquids. But now imagine we perform an experiment on two fluids which are not miscible in the liquid state (i.e., where $T < T_C$ for both fluids). Suppose that we begin by mixing them in the gas state at $T \gg T_C$ for both fluids, and low density. Then we increase pressure isothermally and because $T \gg T_C$ the density increases with no first order phase transition until we eventually reach the melting line. Will the fluids remain mixed, or will the increase in density cause a transition to liquid-like behaviour

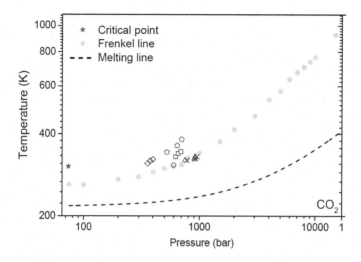

FIGURE 7.5 The phase diagram of CO_2 [6.21] consisting of the experimentally determined melting line (dashed line), critical point (star), and the Frenkel line predicted from MD simulations using the VAF criterion (green circles). Other points are P,T conditions for solubility maxima of different solutes in supercritical CO_2. Reprinted figure with permission from C. Yang, V. V. Brazhkin, M. T. Dove, and K. Trachenko, Phys. Rev. E **91**, 012112 (2015). Copyright (2015) by the American Physical Society.

and a phase separation? It is tempting to jump to the conclusion that a transition to liquid-like mixing behaviour will take place when the Frenkel line for the mixture is crossed, but at the time of writing no-one has properly investigated this matter to the author's knowledge, either experimentally or theoretically. The most relevant work in the literature is some limited conjecture about a link between the Frenkel line and the solubility of some solids in CO_2 [6.21]. In this work, it was noted that the P,T conditions for maximum solubility of certain long-chain molecules (for instance β-carotene) all lie close to, but on the gas-like side of, the predicted Frenkel line for CO_2 (Figure 7.5).

The rationale for this behaviour is the competing effects of density increase: On the one hand, the increase in density increases the dissolving efficiency of the fluid. On the other hand, it results in less (and eventually, no) diffusion of molecules through the fluid, making it harder for the fluid to effectively dissolve the solid. The investigation of this matter is, however, at a very early stage and the transition from gas-like to liquid-like behaviour in miscibility of supercritical fluids is also a topic wide open for future investigation.

References

7.1 A.R. Miedema, P.F. de Châtel and F.R. de Boer, Physica B+C **100**, 1 (1980).

7.2 D.R. Gaskell, *Introduction to the thermodynamics of materials*, Taylor & Francis (2008).

7.3 R. Sander, Atmos. Chem. Phys. **15**, 4399 (2015).

7.4 B.N. Roy, *Fundamentals of classical and statistical thermodynamics*, Wiley (2002).

7.5 E. Detmar, S.Y. Nezhad and U.K. Deiters, Langmuir **33**, 11603 (2017).

7.6 A.F.M. Barton, *CRC handbook of solubility parameters and other cohesion parameters*, CRC Press, Boca Raton (1983).

7.7 J. Delhommelle and P. Millié, Mol. Phys. **99**, 619 (2001).

7.8 M. Waldman and A.T. Hagler, J. Comp. Chem. **14**, 1077 (1993).

7.9 A.E. Nasrabad, R. Laghaei and U.K. Deiters, J. Chem. Phys. **121**, 6423 (2004).

7.10 J. Gernert and R. Span, J. Chem. Therm. **93**, 274 (2016).

7.11 H. Hubel, D.A. Faux, R.B. Jones and D.J. Dunstan, J. Chem. Phys. **124**, 204506 (2006).

7.12 N.W.A. Van Uden, H. Hubel, D.A. Faux, A.C. Tanczos, B. Howlin and D.J. Dunstan, J. Phys.: Cond. Mat. **15**, 1577 (2003).

7.13 P.J. Berryman, D.A. Faux and D.J. Dunstan, Phys. Rev. B **76**, 104303 (2007).

7.14 D.J. Dunstan, N.W.A. Van Uden and G.J. Ackland, High Press. Res. **22**, 773 (2002).

7.15 J.R. Wood et al., Phys. Rev. B **62**, 7571 (2000).

7.16 J.E. Proctor, M.P. Halsall, A. Ghandour and D.J. Dunstan, J. Phys. Chem. Sol. **67**, 2468 (2006).

7.17 C.G. Pruteanu, G.J. Ackland, W.C.K. Poon and J.S. Loveday, Sci. Adv. **3**, e1700240 (2017).

7.18 C.G. Pruteanu, D. Marenduzzo and J.S. Loveday, J. Phys. Chem. B **123**, 8091 (2019).

8

Applications of
Supercritical Fluids

Clearly the applications of liquids in industry are too widespread to review in a single chapter. We will therefore concentrate here on the slightly narrower topic of established and emerging industrial applications of supercritical fluids, covering applications of liquids only where they overlap with those of supercritical fluids. We will cover the applications in power generation, followed by food processing and then other applications.

8.1 Applications of Supercritical Fluids in Power Generation Cycles

8.1.1 Efficiency of Thermodynamic Cycles

It is useful to begin our discussion of the use of supercritical fluids in power generation with some revision of elementary thermodynamics. The efficiency f of the ideal (Carnot) cycle for conversion of heat into work, or any real (e.g., Rankine) cycle is linearly dependent on the ratio of the high (T_H) and low (T_L) temperature endpoints of the cycle (for instance, equation 8.1 for the Carnot cycle). With the low temperature endpoint generally fixed at 300 K or above, a higher temperature endpoint T_H always ensures greater efficiency. This is also the case for real cycles, such as the Rankine cycle. In this case, increasing the range of pressures across which the cycle operates also ensures greater efficiency [8.1].

$$f = 1 - \frac{T_L}{T_H} \qquad (8.1)$$

Therefore, one basic motivation for the use of supercritical fluids in power generation is the desire to push the working fluid to the highest temperature possible, which frequently results in the fluid exceeding the critical temperature and critical density.

8.1.2 Use of Supercritical H₂O in Power Generation

Power cycles utilizing H_2O at supercritical temperatures ($T_C = 647$ K) have been in use in power stations for decades, though generally with the supercritical temperature part of the cycle at significantly subcritical density. A representative example is the H_2O Brayton cycle shown in Figure 8.1 (a) (Drax power station (UK) upon commissioning in 1975 [8.2]). More recently, supercritical H_2O has been used for cycles in nuclear power stations in which significantly supercritical temperatures are achieved at supercritical density. Figure 8.1 (b) shows a representative example of a cycle in a supercritical water nuclear reactor (SCWR) from ref. [8.3] on the V,P phase diagram and Figure 8.1 (c) shows the power cycle from a conventional nuclear reactor (Heysham 1, UK). In this case supercritical temperature is still achieved, but at gas-like density significantly below the critical isochore. Notably, as demonstrated in Figure 8.1 (b) and (c), it is common for significant compression to take place in the mixed phase region. There is therefore a need for EOS that can model this. The accuracy of the fundamental EOS cannot be relied upon in this region.

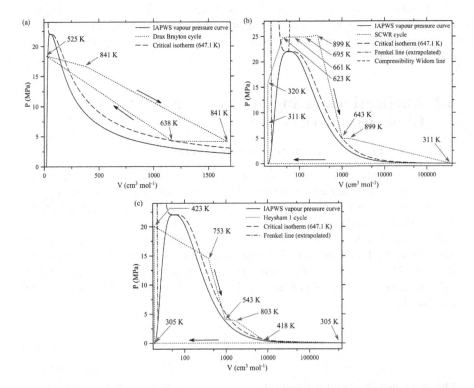

FIGURE 8.1   Example thermal power station cycles plotted on the V,P phase diagram of H_2O alongside the vapour pressure curve plotted from the IAPWS fundamental EOS [6.16]. (a) Drax coal-fired power station Brayton cycle [8.2]. (b) Representative single reheat cycle for supercritical water nuclear reactor [8.3]. (c) Single reheat cycle for conventional water nuclear reactor (Heysham 1, UK).

A stage of the supercritical H_2O cycle crosses the Widom lines, knowledge of which is thus crucial to model these power cycles. Figure 8.1 (b) shows the compressibility Widom line for H_2O, and the C_P Widom line lies very close to this. The power cycles are likely to also cross the Frenkel line, in the $T < T_C$ region. Unfortunately, the Frenkel line has not been directly simulated in the V, P region covered in Figure 8.1. We show instead the author's quadratic extrapolation of the Frenkel line points simulated using the VAF criterion in ref. [6.21]. The lowest pressure simulated was 29.7 MPa so the extrapolation is small and approximately isochoric. The extrapolation terminates on the vapour pressure curve at $0.8T_C$ so is reasonable. We can see in Figures 8.1 (b) and (c) the cycle crossing to the rigid liquid-like side of the Frenkel line at low temperature. In Figure 8.1 (a) the Frenkel line is omitted for clarity.

8.1.3 Use of Supercritical CO_2 in Power Generation

In recent years a massive research and development effort has been initiated into the use of CO_2 as the working fluid in thermodynamic cycles as an alternative to H_2O. The principal application is power generation, but there are also potential applications in the marine sector. The advantages of CO_2 over H_2O are twofold. Firstly, the significantly lower critical point of CO_2 (305 K) makes it possible for a cycle to cover a wide P, T range at accessible temperatures whilst not encountering the first order phase transition across the vapour pressure curve. Secondly, it enables the use of the Allam cycle (ref. [8.4] and refs. therein), a form of Brayton cycle in which combustion of the hydrocarbon fuel takes place in an atmosphere of pure O_2. The combustion product can therefore be the working fluid in the cycle, and the fact that the combustion product is composed virtually entirely of CO_2 and H_2O makes the sequestration and storage of the CO_2 that is produced much easier. The potential benefits are therefore significant and have attracted investment from companies such as General Electric, Toshiba, and Rolls Royce.

Figure 8.2 is a simplified process flow diagram of the Allam cycle demonstrating the key novel feature: The combustion of natural gas in pure O_2 resulting in the combustion product of $CO_2 + H_2O$, in which the H_2O can be easily separated.

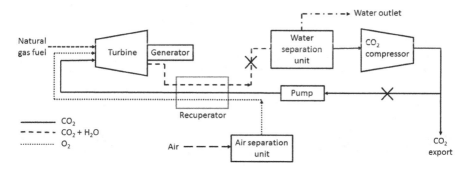

FIGURE 8.2 Schematic diagram showing simplified Allam cycle for power generation by the combustion of natural gas in pure CO_2. The crosses indicate processes to cool the working fluid.

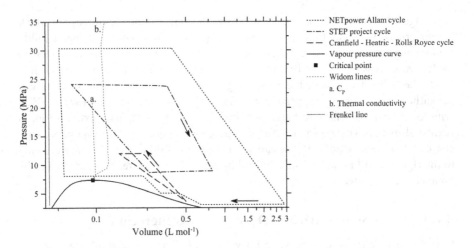

FIGURE 8.3 Supercritical CO_2 power cycles plotted on the V,P phase diagram. V,P points for the NETpower Allam cycle are from ref. [8.4], for the STEP project cycle from ref. [8.5], and the Cranfield–Heatric–Rolls Royce test facility from ref. [8.6]. Data were converted as necessary using the fundamental EOS [3.29] available via NIST REFPROP and the ThermoC code [3.25]. The Widom lines are plotted to their termination using the methodology shown in Section 6.1.1 and the Frenkel line is simulated using the VAF criterion in ref. [6.21].

Figure 8.3 shows three examples of cycles used in these current research and development projects on the V,P phase diagram. We can see, and in the P,T phase diagram Figure 8.4 later, that these cycles cross the Widom lines but do not generally stray close to the Frenkel line. The first cycle is the Allam cycle trialled by NETpower [8.4]. A 50 MWth demonstration plant in La Porte, Texas, began operation in 2019 using this cycle. All CO_2 generated in this power plant is captured. Even without in-built sequestration and storage of CO_2 the use of CO_2 offers significant advantages in terms of efficiency. The second cycle shown in Figures 8.3 and 8.4 is that utilized in the STEP project, a 10 MWe demonstration plant under construction in San Antonio, Texas, at the time of writing. In this cycle the heat generated by conventional fossil fuel burning in air is transferred to CO_2 circulating in the closed-loop cycle shown.

These cycles are also shown on the P,T phase diagram in Figure 8.4. The extremely high temperatures employed in both cases are best for thermodynamic efficiency in power generation. However, they significantly increase operating costs, for instance due to the industry-standard 316L stainless steel failing above ca. 920 K. Specialized high-temperature materials are required for cycles such as the NETpower and STEP projects [8.7]. It is common for smaller-scale research and development projects to utilize cycles operating at significantly lower temperatures [8.8], for instance the test rigs constructed by Cranfield University–Heatric–Rolls Royce collaboration in the UK [8.6] (Figures 8.3 and 8.4) and the test rig at Sandia National Laboratories in the United States [8.9]. As well as being cheaper to operate safely, the lower temperatures utilized in these cycles are

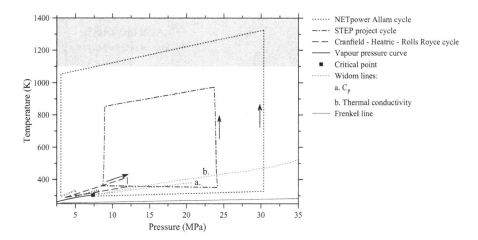

FIGURE 8.4 Supercritical CO_2 power cycles plotted on the P,T phase diagram. V,P points for the NETpower Allam cycle are from ref. [8.4], for the STEP project cycle from ref. [8.5], and the Cranfield–Heatric–Rolls Royce test facility from ref. [8.6]. Data were converted as necessary using the fundamental EOS [3.29] available via NIST REFPROP. The Widom lines are plotted to their termination using the methodology shown in Section 6.1.1 and the Frenkel line is simulated using the VAF criterion in ref. [6.21]. The shaded area indicates the conditions for which it was necessary to extrapolate the fundamental EOS using the ThermoC code [3.25].

the most relevant for some applications such as the bottoming cycle part of combined cycle gas turbine systems. We can observe that these cycles also do not stray close to the Frenkel line, as viewed on either the V,P or P,T phase diagram but do cross the Widom lines at a number of points.

8.1.4 Use of Supercritical N_2 in Power Generation

Supercritical N_2 is also being investigated as a potential working fluid in power cycles, though this research is in its infancy [8.10][8.11]. Some cycles have been investigated theoretically, for instance the Brayton cycle shown in Figure 8.5 proposed by the (South) Korea Atomic Energy Research Institute [8.11]. In this case the N_2 would be heated by the nuclear reactor then would run a turbine in a closed-loop system. N_2 has the benefits of remaining inert, and that machinery developed for use with air can be utilized off-the-shelf with N_2. Notably (in contrast to CO_2) the entire cycle takes place at deeply supercritical temperature but at subcritical density. Neither the Frenkel line nor any Widom lines exist in the P,V,T range covered by Figure 8.5. The Joule-Thomson curve is the only transition existing in this region of the phase diagram. Volumes were obtained using the fundamental EOS [4.3] available via NIST REFPROP, which is valid throughout the P,V,T region shown.

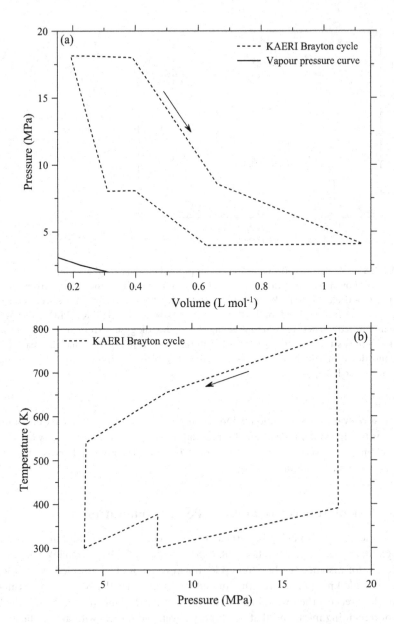

FIGURE 8.5 Proposed supercritical N_2 Brayton cycle by the Korea Atomic Energy Research Institute (KAERI) [8.11] represented on the V, P phase diagram (a) and P, T phase diagram (b). Volumes have been obtained from the given P, T conditions using NIST REFPROP.

8.2 Use of Supercritical Fluids in Food Processing

8.2.1 Decaffeination

The principal application of supercritical fluids in food processing is in the decaffeination of coffee [8.12]. Supercritical CO_2 has been used since the 1970s for coffee decaffeination, replacing other reagents with significantly higher toxicity such as dichloroethane, ethyl acetate, and even benzene(!). In the process, moist CO_2 is passed through the coffee beans at a temperature of 313–353 K and pressure of 12–18 MPa. This corresponds to ~ $1.1 T_C$ and ~ $2 P_C$ for dry CO_2. These conditions lie firmly on the liquid-like side of the critical isochore and Widom lines for (dry) CO_2 but on the gas-like side of the Frenkel line for dry CO_2.

The utility of CO_2 for decaffeination stems from the fact that, whilst caffeine is soluble in moist CO_2 at these conditions, other chemical components of the coffee bean are not. Hence moist CO_2 can be utilized in the supercritical state to remove caffeine from (moist) raw coffee beans. If the process is performed on roasted beans or after grinding, other components of the coffee beans are removed as well as the caffeine. This leads to impairments to the flavour, unless these components are added back into the coffee at a later stage. The decaffeination process has also been adapted to decaffeinate tea [8.13].

Caffeine lovers will be pleased to know that the caffeine removed from the coffee beans does not go to waste; the moist CO_2 is cooled to the liquid state at 298 K, whereupon the caffeine is recovered in an activated carbon filter for use in energy drinks etc.

8.2.2 Other Food Processing Applications

Supercritical CO_2 has a large range of other applications in food processing. It is used to remove fats/oils, cholesterol, and compounds responsible for desirable flavours/aromas from different foods/drinks in a multitude of processes [8.14]. In addition, it is used in processes to fractionate/refine fats and oils. To a great extent, these applications depend on the highly unusual behaviour of the solubility of various solids in CO_2 in the supercritical region close to the critical point [8.15][8.16]. Solubility of a selection of solids in near-critical CO_2 was studied in detail by Schmitt and Reid in 1986 [8.15] and modelled in a semi-empirical manner by Méndez-Santiago and Teja (1999) [8.16]. Figure 8.6 shows example data illustrating the observed effects: A crossover from solubility being nearly independent of temperature to solubility increasing on temperature increase, observed roughly at the critical isochore.

8.3 Supercritical CO_2 Cleaning and Drying

Supercritical fluids (especially CO_2) have a large range of applications in cleaning and drying. Supercritical fluids offer fundamental advantages over liquids in this respect. Firstly, the lack of surface tension facilitates more effective wetting of surfaces and penetration of pores. Secondly, the lack of a phase transition when the supercritical solvent is removed by pressure release above T_C assists in ensuring that

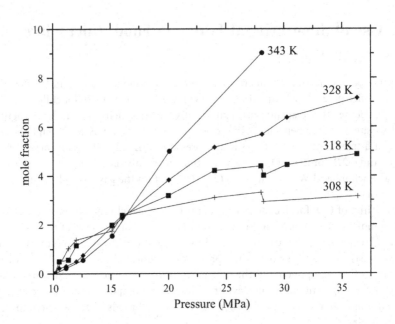

FIGURE 8.6 Solubility of benzoic acid in near-critical CO_2 at various P,T [8.14]. Reprinted from Food Chemistry, **52**, M.V. Palmer and S.S.T. Ting, Applications for supercritical fluid technology in food processing, 345–352, Copyright (1995), with permission from Elsevier.

the process leaves dry surfaces without residuals. Thirdly, the gas-like viscosities combined to some extent with liquid-like densities can lead to very fast cleaning processes.

8.4 Chromatography

Supercritical CO_2 is widely used in chromatography applications as the principal component of the mobile phase. Its utility stems from the combination of relatively high density and low viscosity discussed earlier in relation to other applications. In addition, other solvents with properties competitive with those of supercritical CO_2 have much higher toxicity and are not so readily available.

8.5 Crystal and Nanoparticle Growth

Supercritical fluids (principally H_2O and CO_2) have long had applications in materials synthesis. These stem from the fact that precursor materials can be dissolved in the fluid due to the liquid-like density, whilst the gas-like diffusion speeds ensure that fast reactions are possible.

The growth of solid single crystals from solution is one of the longest-standing applications of supercritical fluids, H_2O in this case. From the end of the 19th century, crystals such as quartz, sapphire, Titanium dioxide, zircon, and tourmaline have been grown from solution in H_2O. The solution is held at typically 20–200 MPa and 573–773 K [8.17] [8.18][8.19][8.20]. These conditions are near or slightly below the critical temperature ($T_C = 647$ K), on the liquid side of the vapour pressure curve when $T < T_C$, but on the gas-like side of the Frenkel line [8.21].

Similar methods have been used to grow a range of nanoparticles from supercritical H_2O and CO_2, for instance TiO_2 [8.19], ZnO, CeO_2, CO_3O_4, various II-VI semiconductor nanoparticles, rare earth vanadates, phosphates, garnets (ref. [8.22] and refs. therein), Pd, Pt and Pd-Pt [8.23], Pt-Ru [8.24], and curcumin (E100 food colouring, also responsible for the yellow colour of natural turmeric) [8.25].

8.6 Exfoliation of Layered Materials

Two-dimensional materials have been one of the hottest topics in physical sciences research over the past two decades. Whilst this field of research did exist prior to 2004, it was the reliable and reproducible isolation of high-quality samples of a single atomic layer of graphite (known as graphene) in 2004 that really kick-started the field [8.26]. The creation of the two-dimensional graphene crystal is made possible by the extreme strength of the (covalent) bonds within the graphene layers compared to the weakness of the (van der Waals) bonds between the layers. Whilst graphite is the most extreme example, many other bulk materials exhibit this property to some extent (BN, MoS_2, WS_2, Ti_3SiC_2, Ti_3AlC_2, etc. [8.27][8.28][8.29]), allowing exfoliation to produce materials one atom or one unit cell thick.

The exfoliation of graphene was originally performed using the now-famous "scotch tape" method of mechanical exfoliation, but this is not really a practical method to produce the large quantities of graphene required for commercial applications. Therefore an enormous variety of other synthesis methods have been tried since 2004. Bottom-up methods have been attempted (epitaxial growth [8.30][8.31], chemical vapour deposition [8.32]), but commercial applications of graphene achieved to date have relied on graphene produced through the top-down method of liquid phase exfoliation [8.27]. In this method, graphite crystals are immersed in a solvent containing a surfactant and subjected to ultrasound. As the graphene sheets are shaken apart using ultrasound, the surfactant molecules attach to the graphene sheets to prevent them from recombining to form graphite. The process is achieved by approximately matching the surface energy of the solvent to that of graphite as required to make the graphene—in effect—soluble in the solvent as described in relation to fluid mixtures in Chapter 7.

This method for producing graphene does not produce samples that are exclusively single atomic layers—flakes of a variety of thicknesses are obtained. When repeated treatment is applied to increase the proportion of monolayers in the sample, damage to the graphene sheets is caused [4.18]. Use of supercritical CO_2 as a solvent has been pursued as a variation on this method, both for graphene [8.33] and other layered materials [8.34]. The use of surfactant is still required to prevent reaggregation to graphite, but in some cases [8.35] it has been possible to dispense with the ultrasound due to the superior dissolving power of supercritical fluids.

References

8.1 J.K. Roberts and A.R. Miller, *Heat and thermodynamics*, Blackie & Son (1960).

8.2 J.G. Collier et al., *Advances in power station construction*, Pergamon Press, Oxford (1986).

8.3 M. Naidin et al., Journal of Engineering for Gas Turbines and Power **131**, 012901 (2009).

8.4 R. Allam et al., Energy Procedia **114**, 5948 (2017).

8.5 https://netl.doe.gov/node/7549

8.6 E. Anselmi et al., *An overview of the Rolls-Royce sCO₂-test rig project at Cranfield University*, proceedings of 2018 sCO₂ power cycle symposium (Pittsburgh, Pennsylvania).

8.7 B.A. Pint and R.G. Brese, *High-temperature materials, in Fundamentals and applications of supercritical carbon dioxide (sCO₂) based power cycles*, Woodhead publishing, Cambridge (2017).

8.8 E.M. Clementoni, T. Held, J. Pasch and J. Moore, *Test facilities, in Fundamentals and applications of supercritical carbon dioxide (sCO₂) based power cycles*, Woodhead publishing, Cambridge (2017).

8.9 S.A. Wright et al., *Summary of the Sandia supercritical CO₂ development program*, proceedings of 2011 sCO₂ power cycle symposium (Boulder, Colorado).

8.10 N. Alpy et al., *Gas cycle testing opportunity with ASTRID, the French SFR prototype*, proceedings of 2011 sCO₂ power cycle symposium (Boulder, Colorado).

8.11 J. Yoon, J. Eoh, H. Kim, D.E. Kim and M. Kim, *Design characteristics of N₂ Brayton cycle power conversion system coupled with an SFR in terms of thermal balance and cycle efficiency*, Transactions of the Korean Nuclear Society Spring Meeting, Jeju, Korea (2018).

8.12 K. Zosel, *Process for the decaffeination of coffee*, USA Patent No. US4260639A (1981).

8.13 M. Gehrig, S. Geyer, J. Schulmeyr, B. Forchhammer and K. Simon, USA Patent No. US7858138B2 (2010).

8.14 M.V. Palmer and S.S.T. Ting, Food Chemistry **52**, 345 (1995).

8.15 W.J. Schmitt and R.C. Reid, J. Chem. Eng. Data **31**, 204 (1986).

8.16 J. Méndez-Santiago and A.S. Teja, Fluid Phase Equilib. **158-160**, 501 (1999).

8.17 E. Kiran, P.G. Debenedetti and C.J. Peters, *Supercritical fluids: Fundamentals and applications*, Springer, New York (2000).

8.18 G. Brunner, *Hydrothermal and supercritical water processes*, Elsevier, Oxford (2014).

8.19 A. Kaleva, S. Heinonen, J.-P. Nikkanen and E. Levänen, IoP Conf. Ser.: Mater. Sci. Eng. **175**, 012034 (2017).

8.20 A. Matthew, American Mineralogist **61**, 419 (1976).

8.21 C. Cockrell, O. Dicks, V.V. Brazhkin and K. Trachenko, https://arxiv.org/abs/1905.00747 (2019).

8.22 K. Byrappa, S. Ohara and T. Adschiri, Adv. Drug Delivery Rev. **60**, 299 (2008).

8.23 B. Cangül, L.C. Zhang, M. Aindow and C. Erkey, J. Supercrit. Fluids **50**, 82 (2009).

8.24 E. Castillejos et al., Chem. Cat. Chem. **4**, 118 (2012).

8.25 Z. Zhao et al., Int. J. Nanomedicine **10**, 3171 (2012).

8.26 K.S. Novoselov et al., Science **306**, 666 (2004).

8.27 V. Nicolosi, M. Chhowalla, M.G. Kanatzidis, M.S. Strano and J.N. Coleman, Science **340**, 1420 (2013).

8.28 M.W. Barsoum and T. El-Raghy, Am. Sci. **89**, 334 (2001).

8.29 J.-P. Palmquist and U. Jansson, Appl. Phys. Lett. **81**, 835 (2002).

8.30 C. Berger et al., Science **312**, 1191 (2006).

8.31 C. Berger et al., J. Phys. Chem. B **108**, 19912 (2004).

8.32 K.S. Kim et al., Nature **457**, 706 (2009).

8.33 N.-W. Pu, C.-A. Wang, Y. Sung, Y.-M. Liu and N.-D. Ger, Mat. Lett. **63**, 1987 (2009).

8.34 Y. Wang, C. Zhou, W. Wang and Y. Zhao, Ind. Eng. Chem. Res. **52**, 4379 (2013).

8.35 E. Lesellier and C. West, J. Chromat. A **1382**, 2 (2015).

9

Supercritical Fluids in Planetary Environments

Helen E. Maynard-Casely

9.1 Introduction

By their very nature of being able to "effuse like a gas, dissolve like a liquid," supercritical fluids can have a profound impact upon their surroundings. However, understanding these impacts has been hampered by the complex and inaccessible nature of the natural environments they are often found in. Table 9.1 outlines a number of regions where supercritical fluids have been observed or inferred to occur. Additionally, the explosion in the numbers of exoplanets that have been discovered, many of them larger and with more extreme environments than found in our own solar system, then the propensity for supercritical fluids to be found in planetary environments is only set to increase. For this chapter, the focus will be exclusively to discuss the impact of supercritical fluids, rather than also including the liquid state which is discussed in the rest of the book. Many treatises have already been undertaken on liquids (principally water) in planetary environments; the role of this chapter is to review the impact of supercritical fluids.

A good example of the difficulties in approaching the study of supercritical fluids is in the understanding of a fundamental process on our own Earth, that of subducting plate margins. Though they have a vital role in the re-processing of continental crust and replenishment of our atmosphere we are yet to sample the fluids that are assumed to govern the movement of subducting slabs. It has long been assumed that supercritical water and potentially carbon dioxide will play a major role in the mechanics and mineral transformations occurring there. However, as we are yet to sample this fluid, and that the effect of a supercritical fluid is highly dependent on impurities, the full understanding of the role of supercritical fluids within these complex environments remains elusive. Hence, laboratory work which feeds into geochemical models and calculations form the basis of our understanding of these inaccessible environments and their wider effects. The first role of this chapter is to explain how supercritical fluids modify mineral interactions and how this has been characterized and analyzed to date.

TABLE 9.1 Locations and typical conditions where supercritical phenomena would be found in planetary environments

Location	Temperature (K)	Pressure (MPa)	Reference
Surface of Venus	~460	9.3	Kuzmin and Marov 1975
Comfortless Cove hydrothermal field	~400	~30	Koschinsky et al. 2008
Terrestrial subduction zones	450 – 1000	500 – 4000	Kohn et al. 2018
Terrestrial inner core boundary	4000 – 8000	320 000	Wijs et al. 1998; Poirier 1991
Ice mantles of Uranus and Neptune	2000 – 8000	$10^6 - 10^8$	Fortney and Nettelmann 2010

With the understanding of the mineral interactions, the way supercritical fluids have influenced the surface and subsurface environments of both terrestrial and icy planetary surfaces can be considered. Noting the environments outlined in Table 9.1, this section will explain the geological results of both small- and large-scale supercritical fluid interactions. Interactions will be examined upon both the terrestrial planets (Mercury, Venus, Earth, and Mars) as well as other solid surfaces such as the icy surfaces of Jupiter's moon Europa, Saturn's moon Titan, and the dwarf planet Pluto. These latter environments are particularly pertinent, where molecular liquids that have not in the past been considered part of a geological system (methane, ethane, and nitrogen) are driving hydrological cycles upon Titan and Pluto.

Perhaps the ultimate application for the understanding of supercritical fluids is upon planetary interior environment, and this is what the last section will address. Planetary interiors are intrinsically high-pressure environments where the majority of fluids will likely be within a supercritical state. As such, these fluids can significantly impact the heat transport within interiors and can hold the balance between conductive and convective regimes.

9.2 Mineral and Material Processes with Supercritical Fluids

In industrial settings, processes that extract selected elements, deposit fine particulates, grow crystallites, and progress chemical reactions with supercritical fluids have been extensively studied and refined so the conditions can be tightly controlled. In planetary environments such tight and optimized control will of course not exist; however, all the same processes will occur. As such the interaction of minerals with supercritical fluids is extremely complex and results in multiple feedback processes as material is dissolved, reaches saturation, and deposited before potentially being re-dissolved. In this section, examples of how these processes from supercritical fluids can occur and how they relate to planetary environments are given. These processes are in addition to the more general phenomena that occur in supercritical fluids that have been addressed elsewhere in this book.

9.2.1 Dissolution of Minerals

If you take a handful of quartz sand and drop this into a glass of water not much will happen, the sand will sink to the bottom and stay there. At atmospheric pressure and room temperature the solubility of quartz (SiO$_2$) in water is almost zero. However, with increasing temperatures and pressures this changes dramatically, an effect that is only complicated by the supercritical regime of water. Nacken first noted the ability of dense steam to dissolve quartz crystals in 1950 (Nacken 1950), and this observation was perhaps the precursor for many of the industrial applications that have been described elsewhere. Latterly, with a greater understanding of thermodynamic variables, it is now possible to calculate solubility over a large range of pressures and temperatures (Shock et al. 1989).

As can be seen in Figure 9.1, at lower pressures a maximum can be observed in the solubility of quartz. This phenomenon can be explained in terms of the sharp change in transport properties that water undergoes when passing through its critical point. This change also results in water becoming very capable of dissolving polar bonds, such as those within quartz and other silicate minerals. However, as can be seen in Figure 9.1, further increases in temperature above 374°C, at pressures less than 600 bar (60 MPa), a decrease in density and dielectric constant of water causes a decrease in the solubility. This is overcome at pressures greater than 750 bar, where the density of the supercritical fluid is enough to maintain an increase in the solubility of quartz with temperature.

The decrease in the ability of water to dissolve quartz at moderate pressures and in the region to 450°C is observed in the interaction with most mineral systems. This is not the case for organic material, which can reach almost total miscibility in the supercritical regime of water (Brunner 2009). Hence any organic material that has co-formed with a mineral system can be separated when the environment includes supercritical water.

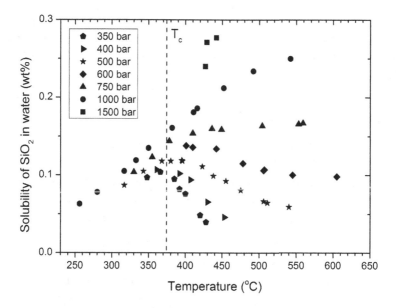

FIGURE 9.1 Changes in the solubility of quartz in water with pressure and temperatures, plotted from experiments by Kennedy (Kennedy 1950). The dotted line indicates the critical temperature (T$_c$) for water.

Pure water will never be found in nature and the interaction with supercritical fluids will always be complicated by dissolved species. The presence of carbon dioxide, for instance, has the effect of lowering the critical temperature but increasing the critical pressure of water. Many studies have investigated water, carbon dioxide, and NaCl mixtures as these are the most common species to comprise geological fluids; a good summary of this work has been written by Duan and Sun (2003). With greater depth comes the potential that the properties of silicate melts and hydrous fluids become less distinct. The increased solubility seen in supercritical fluids could lead to a state where an equilibrium between the silicate melts and hydrous fluids is reached. Measurements of the albite-water system showed this to be the case (Shen and Keppler 1997), suggesting that below a critical depth there is only hydrous melt and that the conditions will not allow for discrete melting events.

9.2.2 Mineral Reactions

As perhaps expected, supercritical fluids will have distinct effects on the surrounding rocks and inevitably lead to changes in the constituent minerals. Such processes, or metasomatism, are common in igneous and metamorphic settings and are too numerous for all to be reviewed here. Though many metasomatic processes will involve fluids within the supercritical state, it is not a necessary condition, as most geological settings have abundant time to allow for the slower kinetics in the liquid state. As such, many discussions of metasomic processes do not specifically state that the fluid will be in the supercritical state. In fact, it is not always very well understood if a particular transformation will occur without the fluid being in the supercritical state.

One ubiquitous mineral that many of us have in our bathrooms is talc, $Mg_3Si_4O_{10}(OH)_2$, (though in that formulation it is often mixed with fragrance and corn starch). Within Earth, talc is an alteration product that can both release and take up water to and from subsurface environments, and as a result has become an important component in the study of fluids at depth in the terrestrial environment. In altering magnesium-rich rocks (such as serpentine, $2Mg_3Si_2O_5(OH)_4$) it can release water as detailed below:

$$2Mg_3Si_2O_5(OH)_4 + 3CO_2 \rightarrow Mg_3Si_4O_{10}(OH)_2 + 3MgCO_3 + 3H_2O$$

Which also results in the formation of magnesite ($MgCO_3$), the significance of which is explained subsequently. Talc can also form from the alteration of dolomite ($CaMg(CO_3)_2$) and Silica (SiO_2); the formation of talc can take up water via the following reaction:

$$3CaMg(CO_3)_2 + 4SiO_2 + H_2O \rightarrow Mg_3Si_4O_{10}(OH)_2 + 3CaCO_3 + 3CO_2$$

This typifies the challenges of approaching the understanding of mineral reactions in planetary environments, as there are multiple feedbacks and influences from mixtures of supercritical fluids. Investigations into the first (serpentine) reaction have accelerated recently, due to the prospect of geological storage of CO_2 captured from the environment (Seifritz 1990). Despite the challenges of this process in the presence of water

(Guthrie 2001), there has been a successful full-scale test of turning supercritical CO_2 into rock conducted in Iceland (Matter et al. 2016).

However, these efforts have highlighted the potential physical effects on the minerals in each environment. Such an effect was measured by Ishida et al. (2012), who showed the distinct difference that supercritical CO_2 had on hydraulic fracturing (determined from acoustic emissions) compared to the liquid. Moreover, these experiments found the cracking of the granite blocks they experimented on to occur at lower pressures and result in features that extended in three dimensions, thus indicating that supercritical fluids have the potential to modify the physical properties as well at the chemical properties of minerals within deep terrestrial environments.

9.2.3 Partition of Elements

As radiometric dating is the main methodology for determining the history of mineral assemblages, it is critically important to understand how elements may be transported within these systems. This is only complicated by the change in transport properties in supercritical fluids.

There are a number of examples of deep-Earth samples that show different elemental signatures from what would be expected. For instance, the island arc basalts, rocks formed on chains of volcanic island that appear at convergent plate margins (like the Ryukyu Islands off the South of Japan). These rocks are enriched in litherophile elements (e.g., strontium, rubidium, and lead) but depleted in high-field strength elements (e.g., zirconium, titanium, and hafnium) compared to their surrounding mid-oceanic ridge basalts (Perfit et al. 1980). Tracing lead in particular, experimental studies have indicated that supercritical fluids can strongly partition this element into high-pressure-temperature hydrous melts (Bureau et al. 2007).

Considerations of the fluid environment must also be made when investigating geothermometers, minerals used to track the thermal history of a rock and its formation. This is often conducted with minerals that contain a trace amount of thorium, such as xenotime and monazite (Pyle et al. 2001). But experimental results have shown that composition of the surroundings can influence thorium partitioning, showing an enrichment of this element in chlorine-rich fluids (Schmidt et al. 2007).

9.3 Supercritical Fluids within Surface and Subsurface Environments

Having outlined the general effects that supercritical fluids have on processes within planetary environments, this and the following sections review specific environments within planetary systems where these fluids are a part. In this section the discussion is mainly on relatively low-pressure environments where water and carbon dioxide, mixed with a few other species, will be the dominant supercritical fluids. For instance, the significance of the water phase diagram upon the subsurface of three of our terrestrial planets is illustrated in Figure 9.2, with a number of transitions occurring under 1000 MPa. Further supercritical phenomena, that occur in a wider range of materials within the deep interiors of our terrestrial and gas giant planets, will be address in the next section.

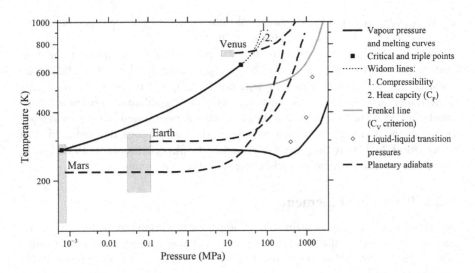

FIGURE 9.2 Figure adapted from (Bargery et al. 2010) indicating the range of surface conditions (grey boxes) and subsurface conditions (planetary adiabats) for three terrestrial planets overlaid on the phase diagram of water, detailing the supercritical phenomena that have been discussed earlier in this book.

9.3.1 Earth

All of the mineral processes discussed so far have been understood from a terrestrial perspective, and that is the lens that we have to understand the other planetary environments. Though Earth's surface and subsurface has much in common with its terrestrial planetary neighbours, being mainly composed of silicate materials, the vital difference is the amount of liquid water to be found here. This means that, on Earth, water dominates the supercritical fluids and shapes the environment accordingly. Our understanding of supercritical fluids on Earth is currently within a transition, as environments where supercritical fluids feature can now be sampled, comparing these with results from prior laboratory work has just begun.

Hydrothermal circulation occurs within the Earth's crust wherever fluid becomes heated and begins to convect and in some settings these fluids are thought to reach supercritical conditions. At undersea plate margins, where the seafloor is thin and allows heat from the interior to warm the ocean, this has formed distinct areas of hydrothermal vents (Figure 9.3). As indicated in Table 9.1, it was recently observed that the ocean itself can reach supercritical conditions in deep sea trenches (Koschinsky et al. 2008). The critical point of sea water has been determined to be 29.8 MPa (below 2940 m) and 680 K (Bischoff and Rosenbauer 1988), compared to the critical point of pure water (22 MPa, 647 K).

The significance of supercritical fluids within these environments is that they would exist within a "two-phase" region of sea water, co-existing with vapour also being produced at elevated temperatures. If conditions of the fluids become critical, then the differences between the vapour and the fluid brine become only minor, which would

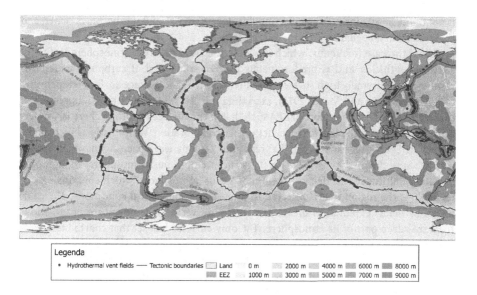

FIGURE 9.3 Distribution of hydrothermal vents on Earth, plotted with data from the InterRidge Vents database version 3.3 (Beaulieu and Szafranski 2018), DeDuijn [CC BY-SA 4.0].

significantly affect metal solubility and fractionation. As a result it has been assumed, though not shown conclusively yet, that hydrothermal vent systems that exceed critical conditions will have very different mineral assemblages and resultant fluids than those that within the sub-critical, two-phase region (Koschinsky et al. 2008).

The Earth is the only planet where active plate tectonics are observed, and this probably correlates with the abundance of liquid water found here. Water can significantly reduce the melting temperatures of rocks within subduction zones, and hence can increase the formation of magma (fluid or semi-fluid material) in these areas. This can then affect both the mobility of the slab being subducted and volcanism in the vicinity of the subducted margin. In these high-pressure/temperature environments any fluid that is liberated from the surrounding rock could then take three forms (or a mixture of each): an aqueous fluid, a hydrous silicate melt, or a supercritical fluid (Mibe et al. 2011). Evidence for the existence of supercritical fluids within these environments is inferred from experimental work investigating the mobility of trace elements (Kessel et al. 2005). The effect of supercritical fluids on these surroundings will be more significant in shallower environments, where conditions are closer to the critical points, and the physical and chemical properties are non-linear (as outlined in Section 9.2.1). At further depths and away from the critical points of water and carbon dioxide there is less variation of properties and complete mixing of many fluid components (Manning 2018).

9.3.2 Other Terrestrial Planets

The lack of water would seem to limit the potential for supercritical fluids to shape the environments of our other terrestrial planets, Mercury, Venus, and Mars. However, this is not the case, and indeed the environments for supercritical fluids becomes richer, and their effects can be planet-wide.

This is the case for Venus, which has a significantly different atmosphere to that of Earth. It is thicker and is predominately (96.5%) comprised of carbon dioxide. This results in an extremely high surface pressure, 9.3 MPa, and average surface temperature of 737 K (Kuzmin and Marov 1975), exceeding the critical temperature and pressure of carbon dioxide (7.39 MPa and 304 K) such that at the surface the atmosphere is a supercritical fluid. From the last few decades of space exploration, with missions such as the European Space Agency's Venus Express and the Japanese Space Agency's Akatsuki, we have very detailed knowledge on the structure of Venus' atmosphere, though only to 7 km of altitude. This deepest section of Venus' atmosphere, which contains 37% of its mass, is the region that is believed to undertake supercritical conditions. The challenges of Venus exploration means that there are very few temperature observations from this deep part of its atmosphere. The only current data set that charted temperature and pressure to the surface of Venus was acquired by the Vega 2 probe (Linkin et al. 1986) Figure 9.4. These data show that, contrary to expectations, the deep atmosphere of Venus is very unstable. Recent modelling to explain this has suggested that the atmospheric stability is affected by density driven separation of nitrogen (the second

FIGURE 9.4   VeGa solar system probe bus and landing apparatus (model) - Udvar-Hazy Center, Dulles International Airport, Chantilly, Virginia, USA image by Daderot [Public domain].

most abundant component of the atmosphere) from carbon dioxide in the supercritical regime (Lebonnois and Schubert 2017).

From radar mapping though the dense clouds it can be seen that lava flows on Venus extend further than their terrestrial equivalents, up to thousands of kilometres. An explanation for this is that the dense supercritical atmosphere does not allow the lava to cool so rapidly as on Earth, and hence keeping it molten for longer allows further flow. The lack of observations also significantly hampers our understanding of how a supercritical atmosphere interacts with the rocks on Venus' surface. A long-time assertion was that the stability of the carbon dioxide in the atmosphere of Venus came from buffering with the surface rocks, though reactions with quartz and calcite (Urey 1952). However, this was latterly shown not to be the case and that transformations of this type would only serve a greater variability in the atmospheric composition of carbon dioxide, but could work to stabilize the smaller gaseous components (Treiman and Bullock 2012). There are no known meteorites from Venus, and our only geochemical data from the surface comes from the Venera 13 and 14, as well as the VeGa 2 landers (Surkov et al. 1984; Surkov et al. 1986) and these data sets have significant uncertainties. They do show that the surface has been heavily altered, and the assumption is that much of this has been from the atmosphere. Investigation into gas-solid reactions has revealed potential alteration pathways (Zolotov 2018), but until we have more geochemical data or a sample returned from Venus its surface will be somewhat of a mystery.

Though Venus is the only planet currently to have a supercritical fluid atmosphere it is likely that other terrestrial planets have also had similarly thick atmospheres at some point in their history. Rapid differentiation on Mars, where water and carbon dioxide out-gassed, may have formed a steam or supercritical atmosphere that would have been present for up to 10^7 years (Lammer et al. 2013). This would have altered the early Martian crust to form a widespread layer of clay minerals. Experimental evidence has supported this (Cannon, Parman, and Mustard 2017), and also suggests that after the thick atmosphere dissipated Mars remained the largely dry planet it is today.

A more localized process that will affect all terrestrial planets is that of impact cratering. Though, on our own Earth, the process of impact cratering is often a neglected consideration in our geological processes, it is clearly a dominant and important one on other terrestrial planets. Investigation of rock alteration on Earth has revealed evidence of changes from supercritical fluids that arise during impact processes (Osinski 2007). The extreme conditions that occur in the formation of these will mean that even silicate material will pass through the supercritical state.

Understanding how these materials respond, and inevitably how the impact processing works, has relied heavily on equations of state. These are still a challenged to predict from theoretical constraints, especially for geologic materials. Hence, most equations of state are fitted from experimental data, and in the case of shock experiments knowledge of the critical point of a material is key to interpreting the results. This is because in most shock experiments the pressure-temperature path will quickly exceed the critical point and it is the pressure release immediately after the impact that allows the material to pass from the supercritical state to form both liquid and vapour. Knowledge of how much of a vapour forms at a given temperature and pressure, as well as how materials may partition between the liquid and vapour phases has greatly influenced our understanding of impact cratering. As the critical point is difficult to assess independently from experiment, empirically modelling has widely been used to assess its pressure and temperature location, as has been done for silica (Melosh 2007). However, with rapidly improving experimental techniques, such as in the use of shock laser pulses (Millot et al. 2015) there is less need to rely on such models to interpret shock experimental data.

9.3.3 Dwarf Planets and Icy Satellites

In addition to the four terrestrial planets and four gas giants in our solar system, we have a number of dwarf planets (Pluto, Ceres, and Eris) and large icy satellites (such as Ganymede, the largest moon of Jupiter, and Titan, the largest moon of Saturn). Water and other volatile species make up a large proportion of these worlds, with silicates and metals buried within their interiors. With, perhaps the exception of Ganymede, where the icy mantle is thought to extend to pressure approaching 1 GPa (1000 MPa) (Vance et al. 2014), it is unlikely that the conditions of these volatile materials will be supercritical (see Figure 9.2). Of course, there will still be the possibility that supercritical fluids will alter the silicate-rich deeper layers of icy satellites and dwarf planets in processes that will be similar to those found in the subsurface and deep seas of Earth.

Icy satellites in our solar system show a variety of features; some are analogous to those observed on terrestrial bodies and some are not. A case in point is in examining impact craters—these are ubiquitous features and are symptomatic of the fact that cratering is one of the most important geological processes. It had been noted that those craters formed on icy moons do not produce features that correspond to their terrestrial planet equivalents. For instance, the gravity of Jupiter's moons Callisto and Ganymede is of the same order as the Earth's moon, but many different morphologies are observed on the icy bodies. Thermal emission measurements of shocks into water ice showed that conditions on impact into ice moved readily into the supercritical regime (Stewart, Seifter, and Obst 2008). Hence, it became important to account for supercritical fluids when modelling icy impact craters. The contrast between this low density phase (the supercritical fluid) and the high-pressure ices that are also shown to be formed are concluded to be the basis for the difference in morphologies seen (Senft and Stewart 2011). However, what effect that any transitions in the supercritical fluid state would have on ice crater morphology is yet to be considered.

9.4 Supercritical Fluids within Planetary Interiors

Aside from that contained in our Sun, most of the mass of the solar system is subject to pressures greater than 10 GPa, that which is locked up under the surface (or clouds) of our planets. As a result, the majority of fluids that occur in the deep interiors of the planets will be within their supercritical regimes. It should be noted that there is an exception to this in that the temperature of liquid core of the Earth is not thought to exceed the critical point of iron, which is about 9250 K and 0.88 GPa (Beutl, Pottlacher, and Jäger 1994). Hence, this section predominately discusses conditions relevant to the interiors of the gas giants as well as speculation on larger terrestrial exoplanets (super Earths). Under these extremes, it is not just molecular materials (water and carbon dioxide for example) that have to be considered within their supercritical regime, but metallic mixtures such as iron and nickel, too. The extremes of these environments, and often the simplicity of the materials at this point, also means that changes in the supercritical state, such as a Frenkel line transition, could have planet-wide implications. Within our solar system there is also a distinction in the chemistry in the gas giants, with Jupiter and Saturn being predominately composed of hydrogen and helium and Uranus and Neptune composed of methane, ammonia, and water. Fundamental advances in the study of these materials as supercritical fluids has gone hand in hand with our increased understanding of the gas giants.

One of the dominant processes that govern planetary interiors is convection, and in this arises an important distinction to be made about terminology. In this chapter (and indeed book) we are concerned about the nature of supercritical fluids. In fluid dynamics the onset of convection, from conduction, is defined when the Rayleigh number of a fluid exceeds its critical value. This is then often referred to the fluid being within a supercritical regime. Though this may indeed also correlate with the particular materials being supercritical fluids, it is not necessarily the case. In this section we will not discuss the supercritical regime that results in convection, rather how the supercritical fluid nature of a material will impact upon planetary interiors.

9.4.1 Jupiter and Saturn

The vast majority of the interiors of Jupiter and Saturn are believed to be hydrogen and helium in their fluid state, as determined from inferred density and measurements of these planets' gravity fields. With the exceptions of the upper atmospheric levels, this fluid will all be at conditions that exceed the critical point of both gases. However, as detailed in Figure 9.5, the conditions within Jupiter and Saturn extend into extremes where these fluids would no longer be classed as supercritical. Instead they can be described as fluids turned metallic by thermal dissociation of the molecules, as discussed for key molecular fluids in Appendix B.

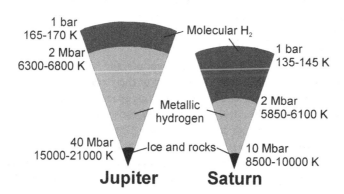

FIGURE 9.5 Schematic illustrating the interiors and transitions within Jupiter and Saturn, boundaries as presented in Fortney and Nettelmann (2010).

One of the most spectacular sites in the solar system is the banded cloud formations of Jupiter. These bands of clouds are observable from even a modest telescope on Earth, but NASA's *Juno* mission's JunoCam (Hansen et al. 2017) has given us all access to the most spectacular views of these cloud features. These bands are associated with flows within the Jupiter that alternate in direction. Known as zonal flows, these are characterized by a fast (~150 ms⁻¹) and wide equatorial jet that flows eastwards, flanked by alternating west/east flowing jets. This phenomenon is not confined to Jupiter, with Saturn exhibiting a similar pattern, though the equatorial jet moves much faster (~300 ms⁻¹). Efforts to understand the mechanisms behind Jupiter and Saturn's zonal flows have centred about two types of models: "weather layer" models, whereby the zonal flows occur in a narrow layer almost confined to the cloud layer, and "deep layer" models, where zonal winds extend through the whole molecular envelope (~10^4 km). The merits of each model were review in the wake of data from the Galileo and Cassini missions (Vasavada and Showman 2005); the deep layer models in particular would be highly influenced by the supercritical fluid state.

One classical problem has been that Saturn's luminosity is greater than what has been estimated from its cooling since formation, suggesting a greater heat flux from the interior of the planet (Hubbard, Guillot, et al. 1999). It has been suggested that there is a region in Saturn where hydrogen and helium become immiscible and the subsequent condensation of helium (helium rain) releases latent heat as it descends deeper in the planet (Stevenson and Salpeter 1977). Investigations into the immiscibility of hydrogen and helium at present rely on theoretical results, but there is a growing consistency in these calculations that such conditions will occur within a large region of Saturn's interior (Morales et al. 2009).

At the time of writing, NASA's *Juno* mission is collecting further detailed measurements of Jupiter's gravity field. In anticipation of these observations, Hubbard and Militzer (2016) were able to use constraints from previous observations and the immiscibility boundary to construct a more detailed model of Jupiter's interior for subsequent comparison. Figure 9.6 plots the lower region of Jupiter's adiabat and its intersection with the immiscibility boundary, which would infer a region of helium rain within Jupiter, though having a more limited affect that the similar region inferred for Saturn.

9.4.2 Uranus and Neptune

A difficulty with the study of Uranus and Neptune, the "icy gas giants" of our solar system, is the lack of observations, principally of their gravitational fields, as neither of these planets have been the subject of an orbiting space mission. Density constraints mean that we can infer their interiors to mostly consist of fluid "ices," a mixture of methane, ice, and ammonia. As detailed in Figure 9.7, the conditions within the majority of these planets are such that we would expect these fluids to be in the supercritical state. Though to what extent they are as intact molecules, or if the adiabat of the planets intersects with a solid superionic phase (as suggested first by Cavazzoni et al. [1999]) has yet to be established.

Like with Saturn, there is a puzzle on the heat flow from Uranus. While current interior models of Neptune can match the observed effective temperature, this is not the

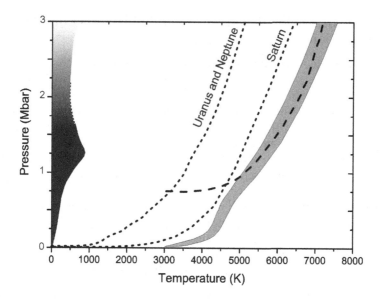

FIGURE 9.6 Grey shaded region plots the calculated adiabat for Jupiter as calculated by Hubbard and Militzer (2016), as well as adiabats indicated for Uranus, Neptune, and Saturn (Guillot and Gautier 2014). All show a possible intersection with the immiscibility boundary for hydrogen (thicker dashed line [Morales et al. 2009]). For reference, black region indicates the solid region of the hydrogen phase diagram.

case for Uranus which would seem to have a much smaller heat flux than expected. There is also the puzzle of their magnetic field generation; both have highly eccentric fields that are not aligned with their cores. There is certainly scope to improve physical understanding of supercritical fluids to contribute to answering these questions.

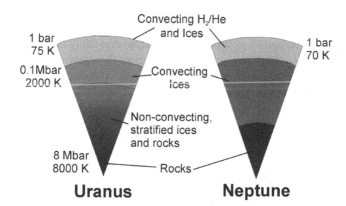

FIGURE 9.7 Schematic illustrating the interiors and transitions within Uranus and Neptune, boundaries as presented in Fortney and Nettelmann (2010).

9.4.3 Transitions in the Supercritical Fluids; Effect on the Gas Giants

As described previously in this book, there has in the last few years been a sizeable shift in our understanding of the physics of supercritical fluids. Of principle relevance to planetary science has been the identification of dynamic transitions in materials such as hydrogen and iron, and that the likelihood that the adiabats of the gas giants will transition through these. This advance in our understanding is also relevant in relation to exoplanets because the range in conditions and chemistry of planets beyond the solar system will only increase the possibilities of such transitions occurring with a wider variety of materials. The current assertion is that there will be a boundary in the properties of supercritical hydrogen around 10 GPa and 3000 K, equating to 4000 and 6000 km below the 1 bar level of Jupiter and Saturn respectively (Trachenko, Brazhkin, and Bolmatov 2014). This has been proposed as a new paradigm for defining the exterior and interior of gas giants and would represent a change from "liquid-like" to "gas-like" dynamics within the supercritical state. The impact that this would have on planetary properties has yet to be fully understood, but work on iron (Fomin et al. 2014) suggests that the transition would also have implication on conductivity which would significantly impact on the generation of magnetic fields. Of particular interest is to understand how a planet-scale transition through the Frenkel would affect heat flow, such as in the case of Uranus explained in the previous section.

9.5 Summary

This chapter has outlined the processes and affects that supercritical fluids can have on planetary environments. This has been undertaken by outlining the potential processes that can be mediated by supercritical fluids, before then discussing the environments where supercritical fluids occur throughout the solar system. This chapter has predominately covered the environment present within our own solar system, but it would be remiss not to comment on the wider possibilities. With the Kepler Space Telescope alone (Koch et al. 2010), 550 terrestrial-candidate planets have now been identified (Morton et al. 2016) as well as many more gas giant type planets. These discoveries have very much adjusted our understanding of how and why planets form, as well as expanding the conditions that we expect to find on and within them. Added to this has been the recent advances in the understanding of the physics of supercritical fluids, as related by this volume. The application of this to planetary science is still very much in its infancy, but as this chapter hopes to relate, has potential to explain some of the enduring mysteries of our solar system and beyond.

References

Bargery, Alistair Simon et al. (2010). "The initial responses of hot liquid water released under low atmospheric pressures: Experimental insights." In: *Icarus* 210.1, pp. 488–506.

Beaulieu, S.E. and K. Szafranski (2018). *InterRidge Global Database of Active Submarine Hydrothermal Vent Fields, version 3.4.* URL: http://vents-data.interridge.org (accessed 5/31/2019).

Beutl, Michael, Gernot Pottlacher and Helmut Jäger (1994). "Thermophysical properties of liquid iron." In: *International journal of thermophysics* 15.6, pp. 1323–1331.

Bischoff, James L. and Robert J. Rosenbauer (1988). "Liquid-vapour relations in the critical region of the system NaCl-H2O from 380 to 415 C: A refined determination of the critical point and two-phase boundary of seawater." In: *Geochimica et Cosmochimica Acta* 52.8, pp. 2121–2126.

Brunner, G. (2009). "Near critical and supercritical water. Part I. Hydrolytic and hydrothermal processes." In: *The Journal of Supercritical Fluids* 47.3, pp. 373–381.

Bureau, H. et al. (2007). "In situ mapping of high-pressure fluids using hydrothermal diamond anvil cells." In: *High Pressure Research* 27.2, pp. 235–247.

Cannon, Kevin M., Stephen W. Parman and John F. Mustard (2017). "Primordial clays on Mars formed beneath a steam or supercritical atmosphere." In: *Nature* 552.7683, pp. 88.

Cavazzoni, C. et al. (1999). "Superionic and metallic states of water and ammonia at giant planet conditions." In: *Science* 283, pp. 44.

Duan, Zhenhao and Rui Sun (2003). "An improved model calculating CO2 solubility in pure water and aqueous NaCl solutions from 273 to 533 K and from 0 to 2000 bar." In: *Chemical geology* 193.3–4, pp. 257–271.

Fomin, Yu D. et al. (2014). "Dynamic transition in supercritical iron." In: *Scientific reports* 4, p. 7194.

Fortney, Jonathan J. and Nadine Nettelmann (2010). "The interior structure, composition, and evolution of giant planets." In: *Space Science Reviews* 152.1–4, pp. 423–447.

Guillot, Tristan and Daniel Gautier (2014). "Giant planets." In: *arXiv preprint arXiv:1405.3752*.

Guthrie, G. et al. (2001). "Geochemical aspects of the Carbonation of Magnesium Silicates in an Aqueous Medium." *Los Alamos National Lab.*, URL: http://lib-www.lanl.gov/la-pubs/00796385.pdf

Hansen, C.J. et al. (2017). "Junocam: Juno's outreach camera." In: *Space Science Reviews* 213.1–4, pp. 475–506.

Hubbard, William B., T. Guillot, et al. (1999). "Comparative evolution of Jupiter and Saturn." In: *Planetary and space science* 47.10–11, pp. 1175–1182.

Hubbard, William B. and B. Militzer (2016). "A preliminary Jupiter model." In: *The Astrophysical Journal* 820.1, p. 80.

Ishida, Tsuyoshi et al. (2012). "Acoustic emission monitoring of hydraulic fracturing laboratory experiment with supercritical and liquid CO2." In: *Geophysical Research Letters* 39.16.

Kennedy, George Clayton (1950). "A portion of the system silica-water." In: *Economic Geology* 45.7, pp. 629–653.

Kessel, Ronit et al. (2005). "Trace element signature of subduction-zone fluids, melts and supercritical liquids at 120–180 km depth." In: *Nature* 437.7059, p. 724.

Koch, David G. et al. (2010). "Kepler mission design, realized photometric performance, and early science." In: *The Astrophysical Journal Letters* 713.2, p. L79.

Kohn, Matthew J. et al. (2018). "Shear heating reconciles thermal models with the metamorphic rock record of subduction." In: *Proceedings of the National Academy of Sciences* 115.46, pp. 11706–11711.

Koschinsky, Andrea et al. (2008). "Hydrothermal venting at pressure-temperature conditions above the critical point of seawater, 5 S on the Mid-Atlantic Ridge." In: *Geology* 36.8, pp. 615–618.

Kuzmin, Arkadi Dmitrievich and M. Ya Marov (1975). *Physics of the planet Venus.* NASA Transl. into English, Nauka Press, 1974. URL: https://archive.org/details/nasa_techdoc_19750018887/mode/2up

Lammer, Helmut et al. (2013). "Outgassing history and escape of the Martian atmosphere and water inventory." In: *Space Science Reviews* 174.1–4, pp. 113–154.

Lebonnois, Sebastien and Gerald Schubert (2017). "The deep atmosphere of Venus and the possible role of density-driven separation of CO 2 and N 2." In: *Nature Geoscience* 10.7, p. 473.

Linkin, V.M. et al. (1986). "Vertical thermal structure in the Venus atmosphere from Provisional VEGA-2 temperature and pressure data." In: *Soviet Astronomy Letters* 12, pp. 40–42.

Manning, Craig E. (2018). "Fluids of the lower crust: deep is different." In: *Annual Review of Earth and Planetary Sciences* 46, pp. 67–97.

Matter, Juerg M. et al. (2016). "Rapid carbon mineralization for permanent disposal of anthropogenic carbon dioxide emissions." In: *Science* 352.6291, pp. 1312–1314.

Melosh, H.J. (2007). "A hydrocode equation of state for SiO2." In: *Meteoritics & Planetary Science* 42.12, pp. 2079–2098.

Mibe, Kenji et al. (2011). "Slab melting versus slab dehydration in subduction-zone magmatism." In: *Proceedings of the National Academy of Sciences* 108.20, pp. 8177–8182.

Millot, M. et al. (2015). "Shock compression of stishovite and melting of silica at planetary interior conditions." In: *Science* 347.6220, pp. 418–420.

Morales, Miguel A. et al. (2009). "Phase separation in hydrogen–helium mixtures at Mbar pressures." In: *Proceedings of the National Academy of Sciences* 106.5, pp. 1324–1329.

Morton, Timothy D. et al. (2016). "False positive probabilities for all Kepler objects of interest: 1284 newly validated planets and 428 likely false positives." In: *The Astrophysical Journal* 822.2, p. 86.

Nacken, R. (1950). "Die hydrothermale mineralsynthese als grundlage zur zuchtung von quarzkristallen." In: *Chemiker-Ztg. 74. Jahrg. Nr.* 50, pp. 745–749.

Osinski, Gordon R. (2007). "Impact metamorphism of CaCO3-bearing sandstones at the Haughton structure, Canada." In: *Meteoritics & Planetary Science* 42.11, pp. 1945–1960.

Perfit, R. et al. (1980). "Chemical characteristics of island-arc basalts: implications for mantle sources." In: *Chemical Geology* 30.3, pp. 227–256.

Poirier, J.P. (1991). "Introduction to physics of the Earth's interior." Cambridge University Press, Cambridge.

Pyle, Joseph M. et al. (2001). "Monazite–xenotime–garnet equilibrium in metapelites and a new monazite–garnet thermometer." In: *Journal of Petrology* 42.11, pp. 2083–2107.

Schmidt, Christian et al. (2007). "In situ synchrotron-radiation XRF study of REE phosphate dissolution in aqueous fluids to 800 C." In: *Lithos* 95.1–2, pp. 87–102.

Seifritz, W. (1990). "CO2 disposal by means of silicates." In: *Nature* 345, p. 486.

Senft, Laurel E. and Sarah T. Stewart (2011). "Modeling the morphological diversity of impact craters on icy satellites." In: *Icarus* 214.1, pp. 67–81.

Shen, Andy H. and Hans Keppler (1997). "Direct observation of complete miscibility in the albite–H2O system." In: *Nature* 385.6618, p. 710.

Shock, Everett L., Harold C. Helgeson and Dimitri A. Sverjensky (1989). "Calculation of the thermodynamic and transport properties of aqueous species at high pressures and temperatures: Standard partial molal properties of inorganic neutral species." In: *Geochimica et Cosmochimica Acta* 53.9, pp. 2157–2183.

Stevenson D.J. and E.E. Salpeter (1977). "The phase diagram and transport properties for hydrogen-helium fluid planets." In: *Astrophysical Journal Supplement Series* 35.10, pp. 221–237.

Stewart, Sarah T., Achim Seifter and Andrew W. Obst (2008). "Shocked H2O ice: Thermal emission measurements and the criteria for phase changes during impact events." In: *Geophysical Research Letters* 35.23.

Surkov, Yu A., V.L. Barsukov et al. (1984). "New data on the composition, structure, and properties of Venus rock obtained by Venera 13 and Venera 14." In: *Journal of Geophysical Research: Solid Earth* 89.S02, pp. B393–B402.

Surkov, Yu A., L.P. Moskalyova et al. (1986). "Venus rock composition at the VeGa 2 landing site." In: *Journal of Geophysical Research: Solid Earth* 91.B13.

Trachenko, K., V.V. Brazhkin and D. Bolmatov (2014). "Dynamic transition of super-critical hydrogen: Defining the boundary between interior and atmosphere in gas giants." In: *Physical Review E* 89.3, p. 032126.

Treiman, Allan H. and Mark A. Bullock (2012). "Mineral reaction buffering of Venus' atmosphere: A thermochemical constraint and implications for Venus-like planets." In: *Icarus* 217.2, pp. 534–541.

Urey, Harold Clayton (1952). "The planets: Their origin and development." Cumberlege, London.

Vance, Steve et al. (2014). "Ganymede's internal structure including thermodynamics of magnesium sulfate oceans in contact with ice." In: *Planetary and Space Science* 96, pp. 62–70.

Vasavada, Ashwin R. and Adam P. Showman (2005). "Jovian atmospheric dynamics: An update after Galileo and Cassini." In: *Reports on Progress in Physics* 68.8, p. 1935.

de Wijs, Gilles A. et al. (1998). "The viscosity of liquid iron at the physical conditions of the Earth's core." In: *Nature* 392.6678, p. 805.

Zolotov, Mikhail Yu (2018). "Gas–solid interactions on Venus and other Solar System bodies." In: *Reviews in Mineralogy and Geochemistry* 84.1, pp. 351–392.

Appendix A: Reference Data on Selected Atomic Fluids

A.1 Table of Phase Change Properties for He, Ne, and Ar

Table A.1 shows principal phase change properties for the atomic fluids examined here. Unless otherwise cited, data are from the relevant fundamental EOS (ref. [A.1] for He, [A.2] for Ne and [4.1] for Ar).

A.2 Phase Diagram of He

He and H_2 are the only fluids examined in this text for which the classical liquid conditions (Section 2.1) are not satisfied throughout the fluid sections of the phase diagram. In fact, quantum effects have a far larger effect on the phase diagram of He than on that of H_2 (which is outlined in Appendix B). In the first instance, the phase diagrams of ^{4}He and ^{3}He differ significantly [4.6]. This is not merely due to the substantial difference in atomic mass; the differing nuclear spin of ^{4}He and ^{3}He plays a key role. Secondly, the critical temperatures of both ^{3}He and ^{4}He are roughly an order of magnitude lower than that for H_2. Therefore the P,T conditions for which the classical fluid conditions are not fulfilled extend much further beyond the critical temperature for ^{4}He and ^{3}He than for H_2.

Here we shall concentrate exclusively on the phase diagram of ^{4}He. Beginning with the Widom lines, referring to Table 6.2 we can observe that He is the only fluid studied here for which the behaviour of the Widom lines deviates significantly from what would be expected due to the principle of corresponding states. Figure A.1 shows the Widom lines for He. In one case (C_P) the Widom line extends to $4.4T_C$, vastly in excess of any of the C_P Widom lines for other fluids terminated according to the same criterion.

In addition, the behaviour of some fluid properties as a function of reduced pressure and temperature (P_R, T_R) is significantly different in He to the behaviour in all other fluids studied in this text. As a result, the criteria outlined in Section 6.1.1 could not be employed to fit the Widom lines for C_V, viscosity η, and thermal conductivity β in He.

TABLE A.1 Properties of fluid He, Ne, and Ar. Properties are obtained from the relevant fundamental EOS [4.1][A.1][A.2] unless otherwise cited.

		Fluid		
	Property	He	Ne	Ar
Critical	Temperature (K)	5.2	44.4918	150.687
	Pressure (MPa)	0.2274	2.6786	4.8630
	Density (mol L^{-1})	17.40	23.882	13.4074
Triple	Temperature (K)	N/A	24.56	83.78
	Pressure (MPa)	N/A	0.043	0.0689
	Liquid density (mol L^{-1})	N/A	62.051	35.458
Phase changes at 300 K	Melting pressure (MPa)	11,500 [A.3]	4802 Simon-Glatzel fit (Figure A.8)	1359
	Fluid density at melting curve (mol L^{-1})	226 (theory) [A.4]	122 [A.5][A.6]	45.25 (240 K)
	Solid density at melting curve (mol L^{-1})	233 (theory) [A.4]	127 [A.6]	47.57 (240 K) [4.2]
Phase changes at 0.1 MPa	Melting temperature (K)	N/A	24.57	83.81
	Fluid density at melting curve (mol L^{-1})	N/A	62.1	35.465
	Solid density at melting curve (mol L^{-1})	N/A	74.6 (6.5 K) [A.7]	40.70 [4.2]
	Saturated vapour temperature (K)	4.2163	27.061	87.178
	Liquid density at vapour pressure curve (mol L^{-1})	31.232	59.850	34.949
	Gas density at vapour pressure curve (mol L^{-1})	4.1306	0.46894	0.14279

The anomalous behaviour of the heat capacities due to quantum effects, combined with the limited P,T range in which high accuracy data is available, precluded the use of the heat capacity criterion to plot the Frenkel line for He. The Frenkel line in He is expected to continue far into the supercritical region in the same manner as for other fluids but to date has only been directly simulated at subcritical temperature, using the VAF criterion [A.8]. Figure A.2 is the P,T phase diagram of fluid He showing the vapour pressure curve and simulated Frenkel line alongside the melting curve, which is discussed later. At extremely low temperatures the transition from the liquid phase (phase I) to the superfluid phase (phase II) [4.6] is also shown.

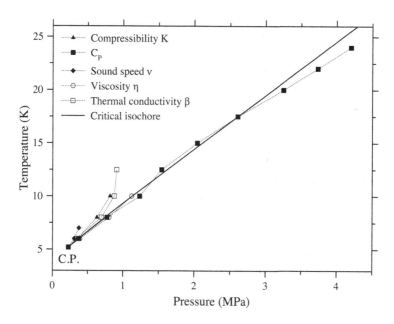

FIGURE A.1 Widom lines and critical isochore for He plotted using the fundamental EOS available from NIST REFPROP.

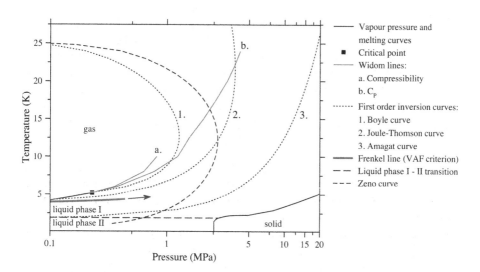

FIGURE A.2 *P,T* phase diagram of fluid He. The vapour pressure curve, Widom line, and inversion curve data are from the fundamental EOS [A.1] obtained using NIST REFPROP and the ThermoC code [3.25], the liquid phase I–II transition line data are from ref. [4.6] and refs. therein, the melting curve data are from refs. [4.6][A.9] and refs. therein, and the simulated Frenkel line data are from ref. [A.8]. The arrow indicates the expected continuation of the Frenkel line.

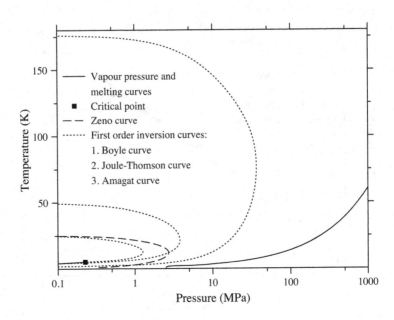

FIGURE A.3   Zeroth order and first order inversion curves for He. Data are sourced from references given in Figure A.2, with additional melting curve data from ref. [A.10]. The Frenkel line, Widom lines, and liquid-liquid transition line are omitted for clarity.

Figure A.2 also shows part of the P,T path taken by the (simulated) zeroth order and first order inversion curves. These curves are shown in their entirety, alongside a larger section of the melting curve, in Figure A.3. The P,T paths taken by the curves are physically realistic. However, even at modest P,T the accuracy of the output from the fundamental EOS [A.1] is not as good as that typically obtained using the fundamental EOS for other fluids. For instance, at $P_R = 2, T_R = 2$ the uncertainty in the P,V,T data is 1% for He [A.1] but only 0.02% for Ar [4.1]. As a result, the potential error in the predicted P,T paths of the inversion curves is significant. We showed examples earlier in this text (Section 6.4.4) where a tiny change in the P,V,T data can have a drastic effect on the path of the inversion curves.

Experimental measurements of the He melting curve are shown up to 1,000 MPa in Figures A.2 and A.3, and the melting pressure at 300 K has been measured accurately (11,500 MPa [A.3]). However, measuring the melting curve at P,T significantly higher than this is very challenging. Due to the tiny atomic volume, He can easily diffuse out of the DAC (usually through the diamonds, breaking them in the process), it has a tiny coherent X-ray scattering cross-section, and no Raman spectrum. Figure A.4 shows a representative selection of the melting curve measurements made on He to date. The curves shown are fits to the experimental data covering the range of the experimental measurements, which lie close to the curve in all cases. The fit parameters are given in Table A.2 (in notation and units consistent with that employed elsewhere in this text). The exception to this is the highest pressure study, the laser heating study of Santamaría-Pérez et al. In this case a significant spread

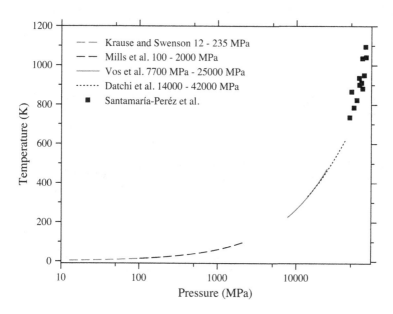

FIGURE A.4 Representative selection of He melting measurements under extreme pressures. Data shown are from Krause and Swenson [A.9], Mills et al. [A.10], Vos et al. [A.12], Datchi et al. [A.13], and Santamaría-Pérez et al. [A.14].

of data around any potential line of best fit is observed, so the individual melting measurements are reproduced in Figure A.4 instead. Melting was detected in these measurements using the laser speckle method. In the case of H_2, the accuracy of this method in detecting the onset of melting has been questioned [A.11]. However, it remains to be seen whether any other method will prove able to detect He melting at such extreme P,T conditions.

TABLE A.2 Fit parameters for He melting curves shown in Figure A.4.

Pressure range (MPa)	Equation	Fit Parameters	Reference
12–235	Glatzel (equation 4.17)	$a = 2.06$ MPa $b = 1.7453$ MPa $c = 1.54681$	Krause and Swenson [A.9]
100–2000	Glatzel (equation 4.17)	$a = 0.8112$ MPa $b = 1.691$ MPa $c = 1.555$	Mills et al. [A.10]
7700–25000	Glatzel (equation 4.17)	$a = 0$ $b = 1.6067$ MPa $c = 1.5650$	Vos et al. [A.12]
14000–42000	Kechin (equation 4.19)	$P_0 = 113.5$ MPa $T_0 = 15.06$ K $a = 125.9$ MPa $b = 1.4988$ $c = -3.9 \times 10^{-6}$ MPa^{-1}	Datchi et al. [A.13]

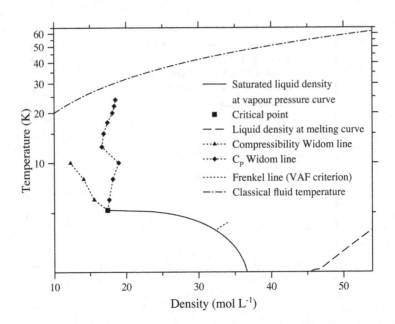

FIGURE A.5 The ρ,T phase diagram of fluid He. The saturated liquid density at the vapour pressure curve is obtained from the fundamental EOS using the ThermoC code [3.25]. The liquid densities at the melting curve are obtained from the fundamental EOS via NIST REFPROP and the Widom line data using the fitting procedure outlined in Chapter 6. The Frenkel line path is obtained from the P,T phase diagram by converting from P to ρ using NIST REFPROP. The classical fluid temperature is plotted as described in the text.

The ρ,T phase diagram of He is presented in Figure A.5. Here, the Frenkel line in the limited range in which it has been simulated using the VAF criterion [A.8] is plotted alongside the saturated liquid density at the vapour pressure curve obtained from the fundamental EOS using the ThermoC code [3.25], and the liquid densities at the melting curve obtained from the fundamental EOS via NIST REFPROP. The uncertainly in these densities is potentially significant. Principal Widom lines (those for the compressibility and C_P) are also shown. The outlier in the C_P Widom line at 10 K is most likely an artefact arising from the fitting process.

We also plot here the temperature at which the classical fluid condition that the spacing between particles is large compared to the de Broglie thermal wavelength is realized. Expressing equation 2.1 in terms of density, we obtain the following equation giving the condition where the spacing between particles is approximately double the de Broglie thermal wavelength:

$$T \approx 4\rho^{\frac{2}{3}} \times \frac{2\pi\hbar^2}{mk_B}$$

In this equation, ρ is measured in molecules per m^3 and m is the atomic mass. Comparing to the relevant graph for H_2 (Figure B.13) we can observe that a lower absolute temperature is required to satisfy the classical fluid condition for He due to the higher atomic mass. However, due to the extremely low critical temperature of He, the region in which the classical fluid conditions are not necessarily fulfilled extends to much higher reduced temperature T_R.

A.3 Phase Diagram of Ne

Ne has a critical temperature (44.49 K) roughly an order of magnitude higher than that of He and much higher mass, ensuring the classical fluid conditions are satisfied throughout. Figure A.6 shows the Widom lines for Ne, alongside the critical isochore, from their beginning at the critical point up to their termination according to the criteria outlined in Section 6.1.1. Data are obtained from the fundamental EOS [A.2] via NIST REFPROP. Unlike the Widom lines in He, the termination of the Widom lines in Ne is in agreement with the principle of corresponding states.

Figure A.7 is the P,T phase diagram of fluid Ne. The vapour pressure curve is obtained from the fundamental EOS [A.2] via NIST REFPROP, as is the Frenkel line plotted using the $c_V = 2k_B$ criterion. The shaded area on the right of the figure indicates (here, and in subsequent figures in Appendices A and B) the (approximate) P,T regions for which the fundamental EOS is not backed by experimental data. In the case of Ne, the heat capacity criterion cannot be used to plot the Frenkel line directly from the c_V data above 225 K, but a reasonable extrapolation of the line is in agreement with the Frenkel line position measured at 300 K using the co-ordination number criterion [4.13].

The fundamental EOS for Ne is not, to the author's knowledge, incorporated into software to plot the inversion curves. The inversion curve data in Figure A.7 are therefore

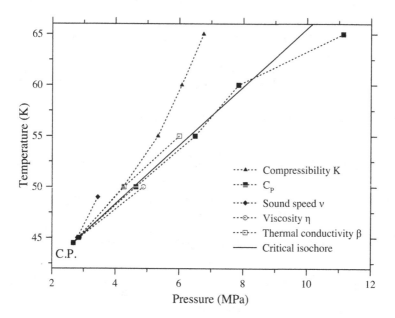

FIGURE A.6 Widom lines for Ne plotted using data from the fundamental EOS obtained via NIST REFPROP and the criteria outlined in Section 6.1.1.

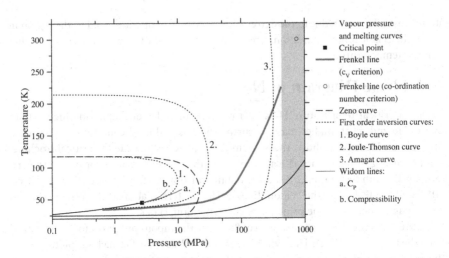

FIGURE A.7 P,T phase diagram of fluid Ne. The vapour pressure curve, Widom lines, and Frenkel line ($c_V = 2k_B$ criterion) are obtained from the fundamental EOS on NIST REFPROP, the Zeno curve and first order inversion curves are obtained from the Xiang-Deiters EOS [A.15] using the ThermoC code [3.25], and the Frenkel line measurement using the co-ordination number criterion at 300 K is from ref. [4.13]. The shaded area indicates the (approximate) P,T region for which the fundamental EOS is not backed by experimental data.

plotted using the Xiang-Deiters EOS [A.15] implemented using the ThermoC code [3.25]. This EOS is also explicit in the Helmholtz function and provides physically realistic paths for the Zeno curve and first order inversion curves. The Amagat curve terminates at ca. 600 K so the high temperature endpoint is omitted from the figure for clarity.

The melting curve of Ne has been studied up to 325 K in several works, principally Crawford and Daniels [A.16] and Vos et al. [A.17]. Figure A.8 shows the P,T datapoints from these two works, fitted with a single Simon-Glatzel equation (4.18) that was also constrained to pass through the triple point. This fit has also been utilized in Figure A.7.

At extreme pressures and temperatures the melting curve of Ne has been studied using the laser speckle method in the laser-heated DAC [A.14] and using Brillouin spectroscopy in the resistively heated DAC [A.18]. The findings from these studies, shown in Figure A.9, are in agreement with each other but not with the Simon-Glatzel fit to the lower pressure data shown in Figure A.8. In any case, there is no indication from the data that the melting curve may exhibit any anomalous behaviour such as a temperature maximum.

Finally, the ρ,T phase diagram of Ne is shown in Figure A.10. The liquid density at the melting curve is omitted since the fundamental EOS is backed by experimental data up to the melting curve only for a very limited P,T range. The saturated liquid density at the vapour pressure curve is shown, along with representative Widom lines and the Frenkel line calculated using the $c_V = 2k_B$ criterion in the P,T range where data are available for this. Notably, the Frenkel line path takes a detour in the region where $T \approx T_C$. This detour is not obviously present when the line is plotted on the P,T phase diagram (Figure A.7), and its origin is not clear since it lies at too high a density to be due to the rise in heat capacities in the vicinity of the critical point.

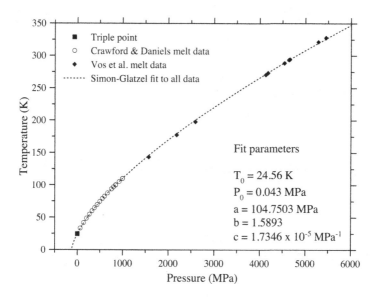

FIGURE A.8 Melting curve of Ne from the triple point [A.2] to 325 K, incorporating the studies of Crawford and Daniels [A.16] and Vos et al. [A.17]. All data are fitted with a single Simon-Glatzel curve (equation 4.18) constrained to pass through the triple point (24.56 K, 0.043 MPa).

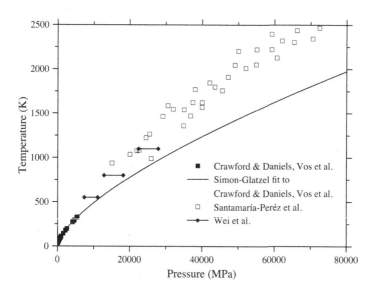

FIGURE A.9 Melting curve of Ne at extreme pressures. Data from Figure A.8 are shown (filled squares) along with the Simon-Glatzel fit to that data. For comparison, the open squares show the measurements made using the laser speckle method in the laser-heated DAC (Santamaría-Pérez et al., ref. [A.14]) and the horizontal lines show the melting curve constraints from the Brillouin spectroscopy study of Wei et al. [A.18] in the resistively heated DAC. The diamonds indicate P,T conditions at which measurements were made identifying whether the sample was in the solid or fluid states, melting therefore takes place somewhere along each of the horizontal lines.

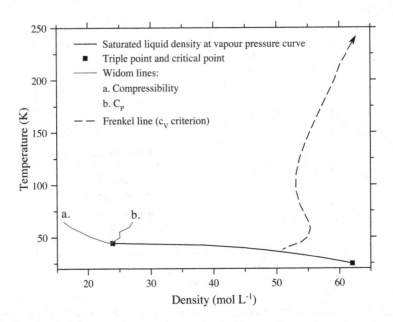

FIGURE A.10   The ρ,T phase diagram of fluid Ne. All data are plotted using the fundamental EOS implemented via NIST REFPROP. The Frenkel line is plotted as far as the fundamental EOS is backed by experimental data, the arrow marks the expected continuation of the Frenkel line.

A.4   Phase Diagram of Ar

Certain aspects of the Ar phase diagram have been outlined in detail in earlier chapters. The Widom line data are shown in Figure 6.5, and data on the P,V EOS and bulk modulus close to the melting curve up to 240 K are shown in Section 4.1.1. Unfortunately, adequate data are not available to examine the P,V EOS and bulk modulus of fluid He and Ar close to the melting curve at ~ 300 K.

The main P,T phase diagram of Ar is given in Figure A.11. A representative selection of the Widom lines are reproduced here, alongside the Zeno curve and first order inversion curves (obtained from the fundamental EOS using the ThermoC code [3.25]) and the Frenkel line plotted from the fundamental EOS using the $c_V = 2k_B$ criterion. In this case, the expected extension of the Frenkel line into the subcritical region cannot be plotted using the heat capacity criterion as it is obscured by the broad maximum in heat capacities occurring in the vicinity of the critical point, as illustrated earlier in Figure 6.17. The melting curve shown in Figure A.12 is obtained from ref. [4.53] as outlined in Section 4.5 and shown in Figure 4.11 (a). Briefly, it is a Simon-Glatzel fit to the available melting data (including that up to 6000 MPa, 800 K obtained in the resistively heated DAC) constrained to pass through the triple point.

Figure A.12 is the ρ,T phase diagram of fluid Ar. The liquid densities at the vapour pressure curve and melting curve are shown, along with the Frenkel line and principal Widom lines. The data are from the same sources as those in Figure A.11.

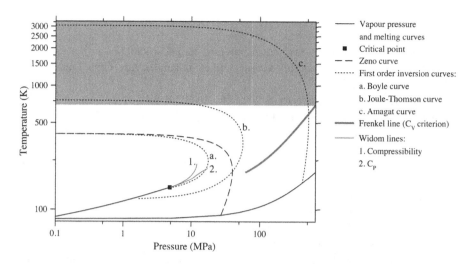

FIGURE A.11 P,T phase diagram of fluid Ar. The vapour pressure curve, Widom lines, and Frenkel line ($c_V = 2k_B$ criterion) are obtained from the fundamental EOS via NIST REFPROP, the Zeno curve and first order inversion curves are obtained from the fundamental EOS using the ThermoC code [3.25]. The shaded area indicates the (approximate) P,T region for which the fundamental EOS is not backed by experimental data.

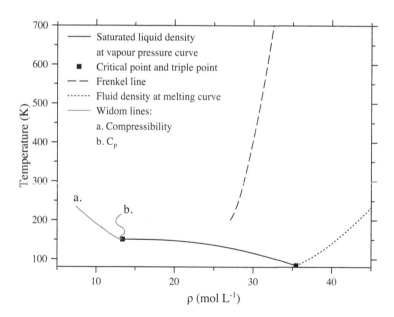

FIGURE A.12 The ρ,T phase diagram of fluid Ar. Data are sourced from the same references as Figure A.11.

References

A.1 R.D. McCarty and V.D. Arp, Adv. Cryo. Eng. **35**, 1465 (1990).

A.2 R.S. Katti, R.T. Jacobsen, R.B. Stewart and M. Jahangiri, Adv. Cryo. Eng. **31**, 1189 (1986).

A.3 J.M. Besson and J.P. Pinceaux, Science **206**, 1073 (1979).

A.4 D.A. Young, A.K. McMahan and M. Ross, Phys. Rev. B **24**, 5119 (1991).

A.5 A. Dewaele, J.H. Eggert, P. Loubeyre and R. Le Toullec, Phys. Rev. B **67**, 094112 (2003).

A.6 L.W. Finger, R.M. Hazen, G. Zou, H.-K. Mao and P.M. Bell, Appl. Phys. Lett. **39**, 892 (1981).

A.7 Y. Endoh, G. Shirane and J. Skalyo Jr., Phys. Rev. B **11**, 1681 (1975).

A.8 A. Takemoto and K. Kinugawa, J. Chem. Phys. **149**, 204504 (2018).

A.9 J.K. Krause and C.A. Swenson, Cryogenics **16**, 413 (1976).

A.10 R.L. Mills, D.H. Liebenberg and J.C. Bronson, Phys. Rev. B **21**, 5137 (1980).

A.11 A. Goncharov, R.J. Hemley and E. Gregoryanz, Phys. Rev. Lett. **102**, 149601 (2009).

A.12 W.L. Vos, M.G.E. van Hinsberg and J.A. Schouten, Phys. Rev. B **42**, 6106 (1990).

A.13 F. Datchi, P. Loubeyre and R. LeToullec, Phys. Rev. B **61**, 6535 (2000).

A.14 D. Santamaría-Pérez, G.D. Mukherjee, B. Schwager and R. Boehler, Phys. Rev. B **81**, 214101 (2010).

A.15 H.W. Xiang and U.K. Deiters, Chem. Eng. Sci. **63**, 1490 (2008).

A.16 R.K. Crawford and W.B. Daniels, J. Chem. Phys. **55**, 5651 (1971).

A.17 W.L. Vos, J.A. Schouten, D.A. Young and M. Ross, J. Chem. Phys. **94**, 3835 (1991).

A.18 W. Wei, X. Li, N. Sun, S.N. Tkachev and Z. Mao, American Mineralogist **104**, 1650 (2019).

Appendix B: Reference Data on Selected Molecular Fluids

B.1 Table of Phase Change Properties for CH_4, CO_2, H_2, H_2O, and N_2

Table B.1 shows principal phase change properties for the molecular fluids examined here. Unless otherwise cited, data for CH_4 are from the Wagner-Setzmann fundamental EOS [3.16], data for CO_2 are from the Span-Wagner fundamental EOS [3.29], data for H_2 are from the fundamental EOS for normal H_2 (a 3:1 mixture of ortho- and para- H_2) from Leachman et al. [B.1], data for H_2O are from the IAPWS (International Association for the Properties of Water and Steam) fundamental EOS [6.16], and data for N_2 are from the Span-Wagner fundamental EOS [4.3].

B.2 Phase Diagram of CH_4

Table B.1 shows key properties of CH_4. The fluid volume upon melting at 300 K was obtained by a minor extrapolation of the P,V relation provided by the fundamental EOS; from 1076 MPa to the measured melting point of 1382 MPa. The values of K_0 and K_0' given in Table B.1 are at 250 K, the highest temperature at which the fundamental EOS is backed by experimental data right up to the melting curve. At 300 K CH_4 solidifies into the face-centered cubic phase I [B.3] with four molecules in the unit cell. The solid density at melting was obtained using a Murnaghan fit to the tabulated X-ray diffraction measurements in ref. [B.3] of phase I, covering the range 1610–5210 MPa.

The orientational order (or lack thereof) in the fluid and solid phases is an interesting topic that may merit renewed investigation. The CH_4 molecules in the solid phase I are believed to not be orientationally ordered ([B.11] and references therein). On the other hand, it is claimed [B.12] that orientational order is observed through X-ray diffraction experiments on liquid CH_4 cooled to 92 K and ca. 0.01 MPa (close to the triple point). It seems highly unusual that orientational order should appear in the liquid state

TABLE B.1 Fluid properties, obtained from the relevant fundamental EOS unless otherwise stated. Numbers marked with an "e" are obtained by an extrapolation as described in the text

Property		Fluid				
		CH_4	CO_2	H_2	H_2O	N_2
Critical	Temperature (K)	190.6	304.1282	33.145	647.096	126.192
	Pressure (MPa)	4.61	7.3773	1.2964	22.0640	3.3958
	Density (mol L^{-1})	10.1	10.6249	15.508	17.873728	11.1839
Triple	Temperature (K)	90.67	216.592	13.957	273.16	63.151
	Pressure (MPa)	0.01169	0.5180	0.0073580	0.000612	0.01253
	Liquid density (mol L^{-1})	28.142	16.779	38.199	55.497	30.9573
Phase changes at 300 K	Melting pressure (MPa)	1382 [4.17]	531.5 [4.26]	5373	996	2637
	Fluid density at melting curve (mol L^{-1})	37.9e	32.459	119e [B.2]	68.651	46.178 (270 K) [B.7]
	Solid density at melting curve (mol L^{-1})	40.56e [B.3]	38.0e [B.4]	125 [B.5]	75.13 [B.6]	47.1026 (270 K) [B.7]
	Saturated vapour pressure (MPa)	N/A	6.7131	N/A	0.0035369	N/A
	Liquid density at vapour pressure curve (mol L^{-1})	N/A	15.434	N/A	0.00142044	N/A
	Gas density at vapour pressure curve (mol L^{-1})	N/A	6.1028	N/A	55.3149	N/A

TABLE B.1 (*Continued*) Fluid properties, obtained from the relevant fundamental EOS unless otherwise stated. Numbers marked with an "e" are obtained by an extrapolation as described in the text

Property		Fluid				
		CH_4	CO_2	H_2	H_2O	N_2
Phase changes at 0.1 MPa	Melting / sublimation temperature (K)	90.72 (melt)	194.6857 (subl.) (0.101325 MPa) (subl.)	13.9837 (melt)	273.16 (melt)	63.1773 (melt) [B.8]
	Fluid density at melting curve (mol L^{-1})	28.143	N/A	Not known	55.5	30.9621
	Solid density at melting curve (mol L^{-1})	31.4 [B.9]	N/A	N/A	51.06	34.8 (53 K) [B.10]
	Saturated vapour temperature (K)	111.51	N/A	20.21	372.76	77.244
	Liquid density at vapour pressure curve (mol L^{-1})	26.341	N/A	35.25	53.212	28.793
	Gas density at vapour pressure curve (mol L^{-1})	0.11186	N/A	0.633	0.032769	0.16265
EOS parameters near melt curve at 300 K (Murnaghan fit)	Bulk modulus K_0 (GPa)	5.029 (250 K)	3.504	Not known	8.3	8.9 (270 K)
	Pressure derivative of bulk modulus K_0'	4.8 (250 K)	5.7	Not known	6.0	4.4 (270 K)

upon cooling but then disappear upon further cooling into the solid state, despite the increase in density.

Figure B.1 illustrates the Widom lines and critical isochore of CH_4 on the P,T phase diagram, obtained from the fundamental EOS using the methodology outlined in Section 6.1. The Widom line for C_V is excluded since it terminates extremely close to the critical point.

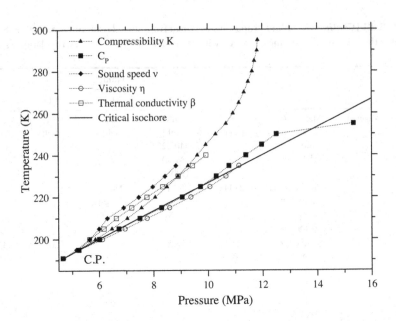

FIGURE B.1 Principal Widom lines and critical isochore of CH_4 shown on the P,T phase diagram. The anomalously high pressure for the C_P Widom line at 255 K is due to a poor fit upon weakening of the maximum due to temperature increase. Lines are plotted using data from the fundamental EOS [3.16] obtained via NIST REFPROP.

In Figure B.2 all the principal transitions in fluid CH_4 are shown. The fundamental EOS is built on P,V,T EOS data up to 625 K and 1000 MPa, as indicated in the figure. Zeroth and first order inversion curves shown were produced using the fundamental EOS via the ThermoC software [3.25]. The high temperature terminations of the Zeno and Boyle curves are thus in the region covered by the fundamental EOS. The Joule-Thomson curve obtained from the fundamental EOS is in agreement with direct measurements of the Joule-Thomson effect performed on CH_4 in the 300–500 K range [B.13] and extrapolates in a reasonable manner to terminate on the temperature axis at ca. 975 K. On the other hand, alternate theoretical/simulation methods to predict the path of the Joule-Thomson curve in the high temperature region produce very different results whilst agreeing with the experimental data up to 500 K [B.13]. The low temperature termination of the Zeno curve is shown, taking place before the melting curve is reached due to the increase in compression factor C as the melting curve is approached.

The Frenkel line shown in Figure B.2 is that predicted from the VAF criterion using MD simulations [6.21]. Optical spectroscopy measurements conducted by the author up to 400 K and by other authors prior to the Frenkel line proposal at 300 K (ref. [4.17] and refs. therein) are in good agreement with the simulation results, considering the substantial experimental error in pressure measurement above 300 K in the DAC. Unfortunately, Frenkel line prediction using the heat capacity criterion is not possible for CH_4 due to the fact that for $T > 200$ K new intramolecular degrees of freedom gradually come into play causing gradual changes to C_V.

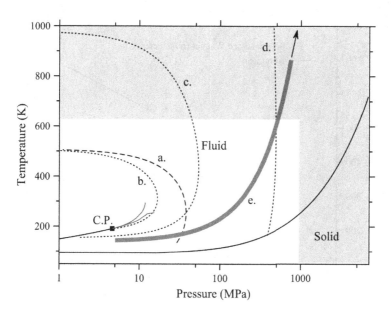

FIGURE B.2 P,T phase diagram of CH_4. Solid black lines indicate first order phase transitions (vapour pressure and melting curves), thin grey lines mark the Widom lines extending furthest from the critical point (compressibility and C_p). The dashed line (a.) marks the Zeno curve and dotted lines mark the first order inversion curves (b. Boyle curve, c. Joule-Thomson curve, d. Amagat curve). The thick grey line marks the simulated Frenkel line [6.21] and the arrow marks the expected continuation of the Frenkel line beyond the P,T range covered in the simulations. The grey shading marks fluid regions for which reliable P,V,T data incorporated into a fundamental EOS is not available.

The phase diagram in Figure B.2 is shown up to 1000 K, since above this temperature the CH_4 molecule begins to thermally decompose (at ambient pressure [B.14]). It is likely that this effect terminates the Amagat curve and Frenkel line for CH_4.

Since the publication of the fundamental EOS for CH_4 in 1991 [3.16] new melting curve measurements have been made at extreme pressures and temperatures in the resistively heated DAC, up to 700 K and 6800 MPa [4.53]. These measurements were combined with selected earlier data and fitted to a Simon-Glatzel curve (equation 4.18, where $P_0 = 1.17 \times 10^{-2}$ MPa, $a = 208$ MPa, $T_0 = 90.6941$ K, $b = 1.698$ for the CH_4 fit) [4.53]. This new fit, which the author considers reliable, departs slightly at high temperatures and pressures from that given in the fundamental EOS. Figure B.3 shows both fits up to the highest P,T at which the fit in ref. [4.53] is backed by experimental data.

Figure B.4 shows the measured melting curve, liquid side of the vapour pressure curve, representative Widom lines (all obtained from the fundamental EOS [3.16]), and the simulated Frenkel line (using VAF criterion, [6.21]) on the ρ,T phase diagram. The lines are shown up to the highest ρ,T for which accurate P,V,T data are available via the fundamental EOS.

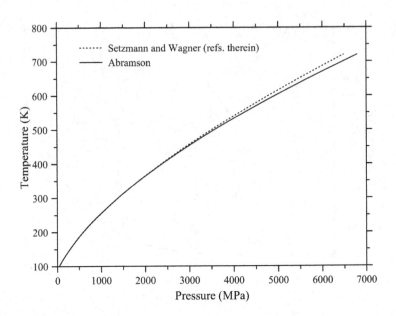

FIGURE B.3 Melting curves given in the fundamental EOS for CH_4 (Setzmann and Wagner, ref. [3.16] and refs. therein) and more recently in Abramson (ref. [4.53] and refs. therein).

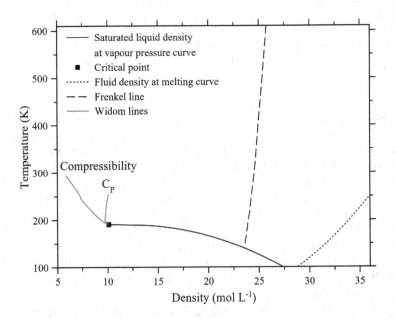

FIGURE B.4 ρ,T phase diagram of CH_4, showing the liquid density at the vapour pressure curve and melting curve, and representative Widom lines all obtained from the fundamental EOS [3.16]. The simulated Frenkel line [6.21] is also shown. The upper ρ,T limits of the figure are the ρ,T for which accurate P,V,T data are available and are incorporated into the fundamental EOS.

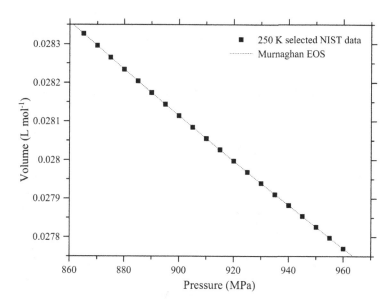

FIGURE B.5 Selected P,V data from the fundamental EOS of CH_4 [3.16] in the vicinity of the melting curve at 250 K (962 MPa) with Murnaghan fit (equation 4.2). Fit parameters are: $K_0 = 5029$ MPa, $K'_0 = 4.8$.

Unfortunately, as shown in Figure B.4 the P,V,T data only reach the melting curve up to 250 K. Therefore in Figure B.5 we show, at 250 K, selected P,V datapoints from the fundamental EOS in the vicinity of the melting curve together with a Murnaghan fit (equation 4.2). This yields the fit parameters shown in Table B.1 for CH_4.

B.3 Phase Diagram of CO_2

Figure B.6 shows the Widom lines of CO_2 plotted from the fundamental EOS and Figure B.7 shows all principal transitions in fluid CO_2, plotted on the P,T phase diagram. The data are obtained from the fundamental EOS for CO_2 [3.29] obtained via NIST REFPROP and the ThermoC software [3.25]. Most of the Joule-Thomson curve can be produced without extrapolation of the fundamental EOS beyond the P,T at which it is backed by experimental data, and the limited extrapolation required does produce a physically reasonable high temperature endpoint. The Amagat curve produced by the fundamental EOS, on the other hand, is unlikely to be reliable. Data are shown in Figure B.7 up to 1400 K. From 1800 K onwards thermal decomposition of CO_2 has been observed in laser heating experiments in the DAC [B.15] and this phenomenon most likely terminates the Amagat line and Frenkel line in CO_2.

The melting curve outlined in the fundamental EOS for CO_2 [3.29] is based on data only up to 270 K. However, since the publication of the fundamental EOS in 1996, a number of DAC studies have extended knowledge of the melting curve

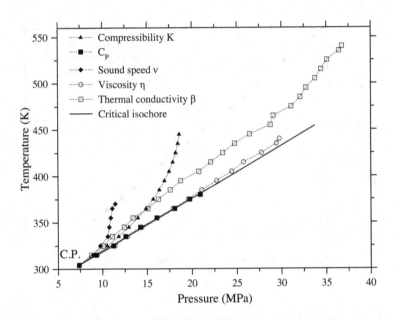

FIGURE B.6 Widom lines of CO_2 plotted from the fundamental EOS via NIST REFPROP on the P,T phase diagram.

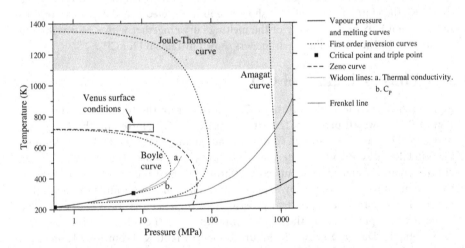

FIGURE B.7 P,T phase diagram of fluid CO_2. Data are obtained from the fundamental EOS [3.29], with the exceptions of the melting curve which is obtained from ref. [4.26] and the Frenkel line, which is simulated using the VAF criterion in ref. [6.21]. The shaded areas mark the P,T limits of the data backing the fundamental EOS. Data are shown up to the highest pressure at which the Frenkel line has been simulated. The black rectangle indicates the range of P,T conditions on the surface of Venus, as discussed in Chapter 9.

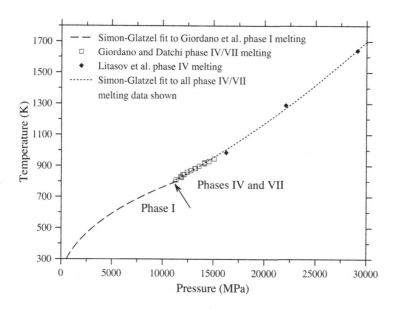

FIGURE B.8 CO_2 melting data from 300 K to the termination of the melting curve due to thermal decomposition of CO_2. Data are compiled from refs. [4.26][B.15][B.17]. Simon-Glatzel fits are shown in Table B.2. The arrow marks the triple point between the fluid state and solid phases I and IV/VII.

right up to the conditions where CO_2 begins to thermally decompose (Figure B.8). Giordano et al. [4.26] studied the melting curve using visual observation and Raman spectroscopy in the resistively heated DAC from 300 K up to 800 K. Their data were combined with data at $T < 300$ K from earlier studies and fitted with a Simon-Glatzel curve (equation 4.18) constrained to pass through the triple point. Their fit is very close to the melting curve given in the fundamental EOS in the temperature region in which they overlap. A separate study in the resistively heated DAC by Abramson [B.16] from 300 K to 673 K obtained similar results (a discrepancy of less than 10 K in the measured melting temperatures). The melting curves from the fundamental EOS [3.29] and ref. [B.16] are therefore omitted from Figure B.8 for clarity.

At ca. 800 K there appears to be a kink in the melting curve, which is expected as it passes through a triple point between the fluid state and two solid phases. On the low pressure side there is phase I, then on the high pressure side phase IV and the metastable phase VII. A separate resistive heating study (Giordano and Datchi [B.17]) focussed on the melting of phases IV and VII from 800–950 K. Melting curve measurements were extended to 1950 K in a study in the laser-heated DAC (Litasov et al. [B.15]), at which point thermal decomposition of the CO_2 molecule was observed. Three direct observations of melting were made, and the melting curve was constrained by a large number of observations of CO_2 in the fluid and solid states at different P,T conditions.

TABLE B.2 Fit parameters from the Simon-Glatzel equation (equation 4.18) fitted to the melting curve of CO_2 solid phase I [4.26] and of solid phases IV and VII (author's fit to data from refs. [B.15][B.17]). P_0, T_0 for phase I fit are set to ensure curve passes through triple point, P_0, T_0 for phase IV/VII fit are adjustable fitting parameters

Phase	P_0 (MPa)	T_0 (K)	a (MPa)	b
I	0.518	216.59	403	2.58
IV / VII	10978.57	799.0765	6.35×10^4	0.347329

In Figure B.8 and Table B.2 a single Simon-Glatzel curve is fitted to selected data points from Giordano and Datchi [B.17] and the three melting data points from Litasov et al. The agreement is good. However, the resulting melting curve is slightly concave towards the temperature axis in the phase IV melting region. Whilst concavity towards the temperature axis can exist in the form of a discontinuous "kink" at the triple point between the fluid and two solid phases, it is extremely rare (Section 4.4). Arguably, CO_2 could be an exception due to the effect of thermal dissociation which could begin to play a role at the temperatures covered in Figure B.8, or a small systematic discrepancy due to the different temperature measurement methods employed in the different studies could be responsible.

At high density in the vicinity of the melting curve, fluid CO_2 has been found to exhibit local structure with orientational order up to 700 K, the highest temperatures investigated to date [B.18]. This order, however, appears gradually rather than in a narrow phase transition. The P,V EOS of solid CO_2 has been studied at 300 K [B.4] and extrapolation of this EOS back to the melting curve suggests a density increase of ca. 20% upon freezing. In the liquid phase at 300 K the bulk modulus and pressure derivative given in Table B.1 were calculated using fundamental EOS data from 350 MPa to freezing at 518.5 MPa.

Figure B.9 shows the saturated liquid density at the vapour pressure curve, liquid density at the melting curve, representative Widom lines, and the Frenkel line for CO_2 on the ρ,T phase diagram. Data are obtained from the fundamental EOS [3.29] via NIST REFPROP and the ThermoC software [3.25], with the exception of the Frenkel line, which is simulated using the VAF criterion in ref. [6.21]. Frenkel line densities are obtained from the pressure via the fundamental EOS. Both pressure and density were monitored during the MD simulations producing the Frenkel line for CO_2 [6.21] and were in agreement with the fundamental EOS to ±5% in the P,T region covered by the fundamental EOS. Data in Figure B.9 are shown up to the highest temperature at which the Frenkel line was simulated in ref. [6.21].

B.4 Phase Diagram of H_2

The phase diagram of H_2 is significantly affected by quantum effects and choice of isotope. In this section we will review the phase diagram of "normal" H_2, a 3:1 mixture of ortho-H_2 and para-H_2. The phase diagrams of pure ortho-H_2, pure para-H_2, D_2 (Deuterium) and T_2 (Tritium) are all different from the diagrams presented here.

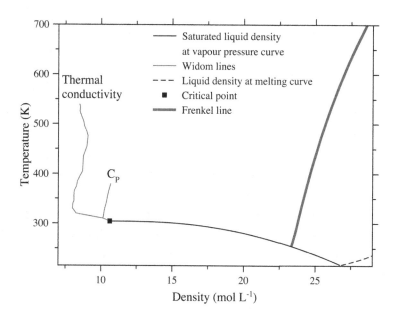

FIGURE B.9 The ρ, T phase diagram of CO_2 showing the saturated liquid density at the vapour pressure curve, liquid density at the melting curve, and representative Widom lines obtained from the fundamental EOS. Thermal conductivity Widom line data have been smoothed. The Frenkel line shown is calculated using the VAF criterion in ref. [6.21].

A fundamental EOS is available for all of these systems [B.1][B.19] except for Tritium for which more limited progress has been made [B.20]. H_2 is also one of the few fluids considered in this text for which the classical fluid condition outlined in equation 2.1 is not satisfied throughout the fluid phase. For the classical fluid condition to be satisfied, the distance between particles must be large compared to the particle de Broglie wavelength, which is temperature dependent. The line marking this criterion is therefore naturally plotted as a density-dependent temperature and is shown later on the ρ, T phase diagram (Figure B.13).

Figure B.10 shows the Widom lines of H_2 plotted out on the P, T phase diagram using the output from the fundamental EOS [B.1]. See Figure 6.7 (b) for an enlargement of the area close to the critical point.

Figure B.11 shows the main P, T phase diagram of H_2. All data are obtained from the fundamental EOS [B.1] via ThermoC [3.25] or NIST REFPROP, except for the melting curve (obtained from Diatschenko et al. [B.21] as discussed below). The fundamental EOS is backed by experimental data up to 1000 K and 2000 MPa, covering the entirety of the fluid part of the phase diagram in Figure B.11.

The Frenkel line in H_2 has not yet been studied in detail. This is unfortunate since H_2 must surely rank as one of the most important systems exhibiting a Frenkel line due to its role in planetary interiors. Its position has been estimated from considerations of the speed of sound and relation to the melting curve on the $\log P, \log T$ phase diagram [B.23], but here we instead utilize the heat capacity criterion. In the temperature range

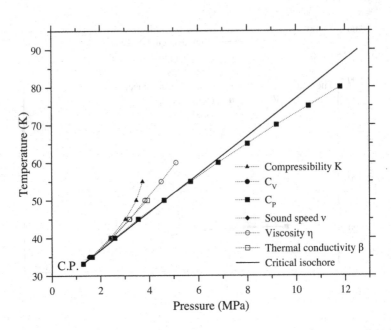

FIGURE B.10 Widom lines of H_2 plotted from the fundamental EOS [B.1] via NIST REFPROP.

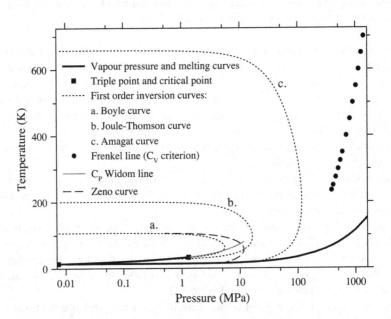

FIGURE B.11 *P,T* phase diagram of fluid H_2 showing the vapour pressure curve, zeroth and first order inversion curves, a representative Widom line (C_P) obtained from the fundamental EOS [B.1] via NIST REFPROP, melting curve (Kechin fit to data from ref. [B.21] as described in text), and Frenkel line points calculated using the C_V criterion as described in the text.

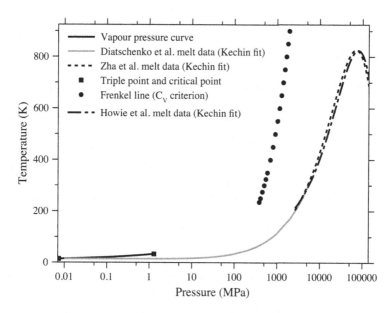

FIGURE B.12 Melting curve of H_2 at extreme pressures. The vapour pressure curve, triple point, and critical point are from the fundamental EOS [B.1], and melting curves shown are Kechin fits to three studies as discussed in the text [B.21][B.24][B.25]. The Frenkel line is calculated using the C_V criterion as outlined in the text.

for which the Frenkel line is shown in Figures B.11 - B.13 the contribution to c_V from intramolecular degrees of freedom remains constant to within ±5% so can be straightforwardly subtracted to allow estimation of the Frenkel line position similarly to noble gas fluids.

The melting curve of H_2 is not considered in the fundamental EOS due to the extreme pressures at which melting takes place (ca. 5.5×10^3 MPa at 300 K). It has, however, been considered in a number of studies in the DAC over the decades. In Figure B.12 the Kechin fits in three key works are shown. Table B.3 shows the fit parameters using our notation (equation 4.19) and units.

TABLE B.3 Kechin fits to H_2 melt P,T points from selected experimental studies. (T.P.) indicates a parameter fixed to co-incide with the triple point, (f) indicates a parameter fixed.

P_0 (MPa)	T_0 (K)	a (MPa)	b	c (MPa^{-1})	Citation
0.0073580 (T.P.)	13.957 (T.P.)	28.4 ± 1.6	1.70 ± 0.03	$-3.5 \times 10^{-6} \pm 4.2 \times 10^{-6}$	Author's fit to data from ref. [B.21]
5500 (f)	300 (f)	7600 ± 400	1.538 ± 0.14	$-8.2 \times 10^{-6} \pm 0.2 \times 10^{-6}$	[B.24] and refs. therein
0 (f)	14.025	29.683	1.66047	-8.591×10^{-6}	[B.25] and refs. therein

We begin with the large number of melting points at temperatures ranging from 25 K ($0.75T_C$) up to 373 K published by Diatschenko et al. in 1985 [B.21]. However, the melting points from this study above 294 K were not reproduced by subsequent studies (see ref. [B.22] and refs. therein in which the melting curve was extended to ca. 800 K, 45×10^3 MPa). In addition, the publication of ref. [B.21] predates the proposal of the Kechin equation and the currently accepted triple point shown in Table B.1. The Kechin fit shown here is therefore the author's own fit disregarding the data above 294 K, but constrained to pass through the accepted triple point. In the Kechin equation as shown in this text, $c < 0$ is necessary for the melting curve to exhibit a temperature maximum at positive pressure. When the Kechin equation is fitted to the data of Diatschenko et al. up to 294 K, $c = 0$ within error and in fact a good fit can also be obtained to the data with $c = 0$ fixed. Thus there is nothing in the melting data up to 300 K to indicate that the melting curve may exhibit a temperature maximum.

On the other hand, in the past five years, two completely independent studies have demonstrated a temperature maximum in the melting curve [B.24][B.25] at extreme pressures of ca. 80×10^3 MPa. In Figure B.12 and Table B.3 the Kechin fits provided in refs. [B.24] and [B.25] are shown. Both are fitted to the melting points in these studies and various references therein. These Kechin fits were not constrained to pass through the triple point, instead the parameters P_0, T_0 were fitted or fixed as listed in Table B.3.

There is thus a consensus from several experimental studies that H_2 exhibits a temperature maximum in its melting curve at ca. 80×10^3 MPa. This is expected theoretically at roughly these conditions [B.26][B.27], a prediction that can ultimately be traced back to the work of Wigner and Huntington in 1935 [B.28]. It is for fundamental quantum mechanical reasons that are outside the scope of this text, as discussed briefly in Section 4.5.1. Whilst dissociation of the H_2 molecule may play a role, this is not due to the amount of thermal energy available. The data in Figure 1.2 shows that dissociation of the H_2 molecule due to thermal energy does not play a role until ca. 3000 K at ambient conditions. At high pressure the temperature required for dissociation due to the presence of thermal energy is expected to increase significantly [B.29].

The experimental works cited above are all studies utilizing resistive heating in the DAC. A number of works have also been published studying the melting curve of H_2 using laser heating in the DAC. It has not been necessary to consider them in the discussion above as resistive heating studies have subsequently become available. Where available, resistive heating is preferred because the errors in pressure and temperature measurement in a laser heating experiment are unavoidably far larger, as outlined in Appendix D.

The ρ, T phase diagram of fluid H_2 is shown in Figure B.13. Here, the saturated liquid density at the vapour pressure curve is shown along with the density at the Frenkel line as plotted earlier using the C_V criterion and the liquid density at the melting curve. Unfortunately there are too few P, V EOS measurements available above 2000 MPa to allow the construction of a fundamental EOS valid up to the melting curve, but there are some limited data available of the P, V EOS as the melting curve is approached, and of fluid volumes at melting at 100–600 K (ref. [B.2] and refs. therein). The fluid densities at

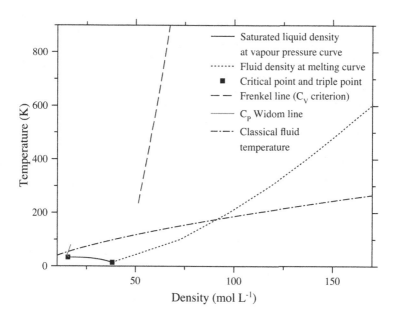

FIGURE B.13 ρ, T phase diagram of H_2. Data for the vapour pressure curve and Widom line are from the fundamental EOS [B.1], the Frenkel line position is calculated from the fundamental EOS using the heat capacity criterion as outlined in the text, and the fluid densities at the melting curve are from ref. [B.2].

melting from ref. [B.2] are shown in Figure B.13 and connected to the triple point given in the fundamental EOS.

We also plot here the temperature at which the classical fluid condition that the spacing between particles is large compared to the de Broglie thermal wavelength is realized. Expressing equation 2.1 in terms of density, we obtain the following equation giving the condition where the spacing between particles is approximately double the de Broglie thermal wavelength:

$$T = 4\rho^{\frac{2}{3}} \times \frac{2\pi\hbar^2}{mk_B}$$

In this expression, ρ is measured in molecules per m³.

Notably, if the Frenkel line calculation using the C_V criterion is to be believed, then the Frenkel line must trend to significantly lower density beneath 235 K in order to terminate (as required) on the vapour pressure curve rather than the melting curve.

Whilst inadequate P, V data are available for fluid H_2 to construct a fundamental EOS valid up to the melting curve at 300 K, an attempt has been made to extrapolate from the available data by Hemmes et al. Shown in Figure B.14 (note the logarithmic axes), this indicates that the compressibility of H_2 decreases enormously from its high value at gas-like pressures. The extrapolation from 2000 MPa up to the melting curve at

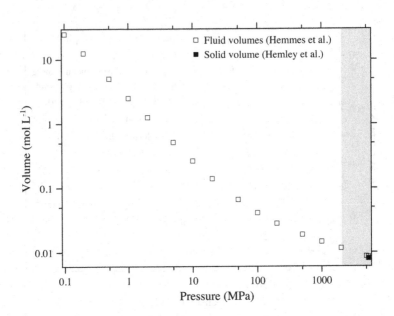

FIGURE B.14   Logarithmic plot of the P,V EOS of fluid H_2 at 300 K. Open squares are output from the EOS fit produced by Hemmes et al. [B.2]. The shaded region ($P > 2000$ MPa) indicates where the fit is extrapolated, rather than backed by experimental data. The filled square indicates the volume of solid H_2 upon freezing measured by Hemley et al. using X-ray diffraction in the DAC [B.5].

5373 MPa may not be accurate enough to justify fitting an EOS to obtain the bulk modulus and pressure derivative in this region as performed elsewhere in this appendix for other fluids. However, it is notable that the extrapolated fluid volume at melting lies close to the solid volume measured accurately using X-ray diffraction in the DAC [B.5].

B.5 Phase Diagram of H_2O

Figure B.15 plots the Widom lines for H_2O in the manner described in Section 6.1.1. The exception is the Widom line for thermal conductivity β, for which (at $T \geq 760$ K) a good fit could not be obtained using this procedure despite the Widom line still clearly being present. To plot the higher temperature part of this line therefore, the maximum values of $(\partial \beta / \partial P)_T$ were measured manually. The Widom line was terminated at the point where we would expect termination due to the corresponding states principle.

Figure B.16 plots two representative Widom lines for H_2O (compressibility and C_P) alongside the zeroth and first order inversion curves, melting curve, and (simulated) Frenkel line on the P,T phase diagram. The inversion curves are plotted from the IAPWS fundamental EOS using the ThermoC software and the melting curve is also that given in the fundamental EOS. This melting curve data is obtained from a number of pre-1970 studies [B.30]. These are in agreement with the more recent study in the resistively heated DAC in the range of overlap shown in this figure (365 K to 440 K) [B.31].

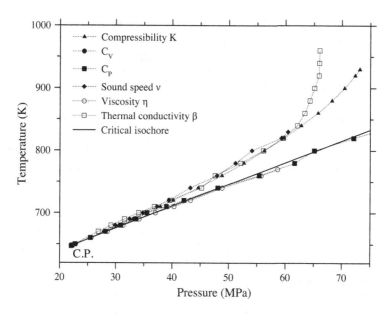

FIGURE B.15 Widom lines of H_2O plotted alongside the critical isochore to their termination. Data are from the IAPWS fundamental EOS [6.16] via NIST REFPROP.

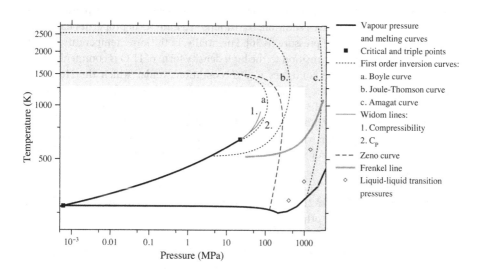

FIGURE B.16 P,T phase diagram of H_2O showing the melting and vapour pressure curves, zeroth and first order inversion curves, representative Widom lines (compressibility and C_p), and Frenkel line.

The P,T range in which the IAPWS fundamental EOS is backed by experimental data is indicated. As we can see, the high temperature terminations of the Zeno, Boyle, and Joule-Thomson curves lie outside this range, but the extrapolation behaviour of the IAPWS EOS does yield physically reasonable paths for these curves. The EOS does not give a physically reasonable path for the high temperature termination of the Amagat curve, as discussed elsewhere [3.28]. Since the H_2O molecule begins to thermally decompose at ca. 2500 K at ambient pressure, and at considerably lower temperature at extreme (multi-GPa) pressures [B.32], the Amagat curve will be terminated by this decomposition in practice. The failure of the IAPWS EOS to predict a physically reasonable termination of the Amagat curve is to be expected given how massive an extrapolation is required beyond the temperature to which it is backed by experimental data.

The Frenkel line shown in Figure B.16 is predicted from MD simulations using the VAF criterion [8.21] and is in agreement with the limited neutron diffraction data available ([8.21] and refs. therein). On the rigid-liquid side of the Frenkel line H_2O is known to exist in two phases (as outlined in Section 4.3.2). The extent to which the transition is narrow enough to justify drawing a line on the phase diagram or is a continuous transition is still debated; nonetheless, attempts have been made to draw out the transition on the phase diagram. The points shown in Figure B.16 were obtained using Raman spectroscopy in the DAC [B.33] and (considering the potential errors in pressure measurement) are in agreement with the results of X-ray [B.34] and neutron scattering studies [4.44][B.35]. On the other hand, at lower temperatures in the vicinity of the melting curve the transition from the low density phase to the high density phase should have a negative slope on the P,T phase diagram due to the negative thermal expansion coefficient of H_2O in this region. In addition, transitions in a number of properties have been noted at 323 K and ambient pressure [B.36]. Potentially, in the lower temperature region covered in this study, the transition to the high density form of H_2O is commencing.

The H_2O melting curve continues to P,T conditions under which the H_2O molecule thermally decomposes. Figure B.17 shows the melting curve from pre-1970 data reviewed by Wagner et al. [B.30] alongside more recent data from the resistively heated DAC (Dubrovinskaia and Dubrovinsky [B.31]) and laser-heated DAC (Goncharov et al. [B.32]), reaching the limit in which thermal decomposition of the H_2O molecule occurs. Whilst interesting and the subject of much debate, this phenomenon is beyond the scope of the present text.

Figure B.18 shows the liquid side of the vapour pressure curve along with the (simulated) Frenkel line [8.21] and representative Widom lines on the ρ,T phase diagram. In common with other fluids, the Frenkel and Widom lines extend to higher temperature roughly along isochores.

In common with other fluids, H_2O demonstrates behaviour similar to a solid in the vicinity of the melting curve. Figure B.19 (a) shows the PV EOS at 300 K from 1 MPa up to the melting curve. The inset shows selected data points from the region close to the melting curve with a Murnaghan EOS fitted (equation 4.2), from which the K_0, K_0' shown in Table B.1 are obtained. What is less usual about H_2O is the thermal EOS at 0.1 MPa, shown in Figure B.19 (b). The negative thermal expansion observed from 273 K to 277 K is highly uncommon, though extending over a small temperature range.

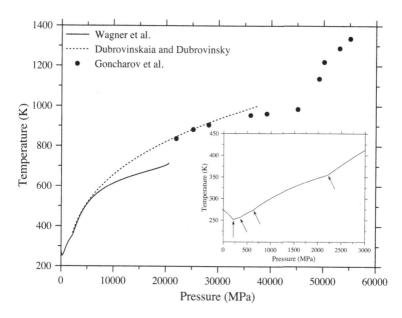

FIGURE B.17 H_2O melting curve data from historic studies (reviewed in Wagner et al. [B.30]), the resistively heated DAC (Dubrovinskaia and Dubrovinsky [B.31]), and laser-heated DAC (Goncharov et al. [B.32]). The inset shows the lower pressure region in which the kinks in the melting curve marked with arrows correspond to triple points between the fluid phase and two different solid phases of H_2O.

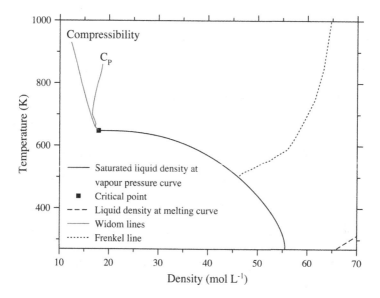

FIGURE B.18 Simulated Frenkel line (from VAF criterion [8.21]), Widom lines and liquid side of the vapour pressure curve (From NIST REFPROP) for H_2O plotted on the ρ, T phase diagram.

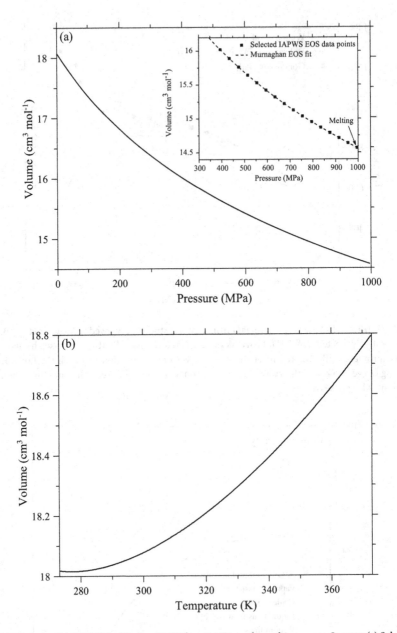

FIGURE B.19 (a) P,V EOS of H_2O at 300 K from 1 MPa to the melting curve. Inset to (a) Selected data points from EOS close to the melting curve, with fit using the Murnaghan equation. (b) Thermal EOS of H_2O at 0.1 MPa from melting to the vapour pressure curve.

B.6 Phase Diagram of N_2

The density and bulk modulus of fluid N_2 close to the melting curve (producing data reproduced in Table B.1) have been discussed in Section 4.1.2. Here we present the Widom lines, Frenkel line, and inversion curves of N_2 on the P,T and ρ,T phase diagrams.

Figure B.20 shows the principal Widom lines of N_2 plotted on the P,T phase diagram using the methodology outlined in Chapter 6 and the output from the fundamental EOS [4.3] via NIST REFPROP. In Figure B.21 the main P,T phase diagram of fluid N_2 is shown. The fundamental EOS is backed by experimental data up to 2200 MPa. In terms of temperature, the fundamental EOS is backed by a large amount of data up to 1000 K and limited data up to 1800 K. Thus the high temperature termination of the Amagat line at ca. 2200 K, whilst looking physically reasonable, does involve a significant extrapolation of the fundamental EOS. The author does not know the origin of the unphysical oscillation in the Amagat curve path at ca. 400 MPa, 200 K.

The vapour pressure curve and inversion curves in Figure B.21 are plotted from the fundamental EOS [4.3] via the ThermoC software [3.25] and NIST REFPROP. The melting curve is from ref. [4.5] and refs. therein, is calculated from the c_V criterion for the temperature range in which this can be used for N_2: At $T < 160$ K the rise in heat capacity due to the proximity of the critical point prevents this criterion from being used to

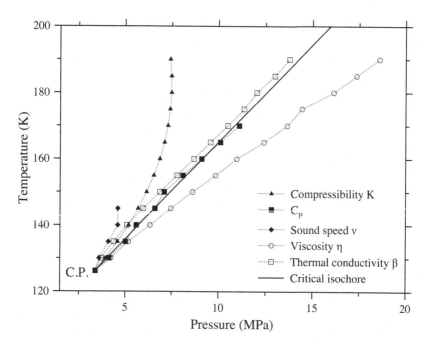

FIGURE B.20 Widom lines of N_2 plotted on the P,T phase diagram utilizing the fundamental EOS [4.3] via NIST REFPROP.

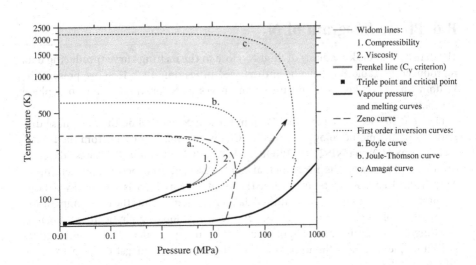

FIGURE B.21   The P, T phase diagram of fluid N_2. The vapour pressure curve, Widom lines, and inversion curves are obtained from the fundamental EOS [4.3] via NIST REFPROP and the ThermoC code [3.25]. The Frenkel line is obtained from the fundamental EOS using the c_V criterion, and the melting curve is obtained from ref. [4.5]. The arrow indicates the expected continuation of the Frenkel line.

obtain the Frenkel line position, and above 315 K the intra-molecular vibrational degree of freedom is gradually activated and affects the heat capacity. The characterization of the Frenkel line in N_2 at 300 K using neutron diffraction, Raman spectroscopy, and MD [4.8] produced a transition pressure agreeing closely with the c_V criterion.

     The melting curve of N_2 was studied at $T < 300$ K by Scott et al. ([4.5] and refs. therein), up to 900 K using Raman spectroscopy in the resistively heated DAC in two separate studies [B.8][B.37] and from 825 K to 2100 K by Weck et al. using synchrotron X-ray diffraction in the laser-heated DAC [B.38]. Figure B.22 summarizes the findings of these studies. The author's Kechin fit to the study in the resistively heated DAC [B.8] is shown, constrained to pass through the triple point. An extrapolation of the Simon-Glatzel fit from Weck et al. (not constrained to pass through the triple point) joins up nicely with the Kechin fit to ref. [B.8] and the data from Mills et al. lies extremely close to the low temperature continuation of this Kechin fit to the triple point. The fit parameters are shown in Figure B.22 using our notation and units (equations 4.18 and 4.19). Fluid N_2 dissociates at temperatures not far above those covered in Figure B.22 [B.39], which is where our interest in the melting curve ends.

     Figure B.23 shows the principal transitions in fluid N_2 on the ρ, T phase diagram, including principal Widom lines, the saturated liquid density at the vapour pressure curve, and the liquid density at the melting curve. The Frenkel line is shown as calculated using the c_V criterion described earlier at the temperatures where this is possible.

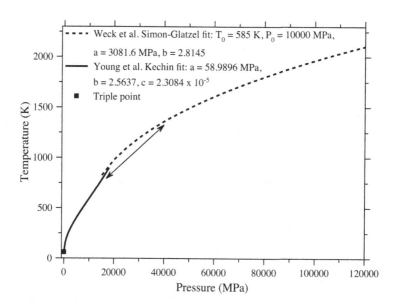

FIGURE B.22 Melting curve of N_2 from the triple point up to the temperature at which thermal dissociation of the N_2 molecule begins [B.39]. Parameters are shown for the author's Kechin fit to the data of Young et al. [B.8] (constrained to pass through the critical point with parameters as shown in Table B.1) and the Simon-Glatzel fit given in Weck et al. [B.38] (not constrained to pass through the triple point). The arrow marks the extrapolation of the fit from Weck et al. performed to link to Young et al.

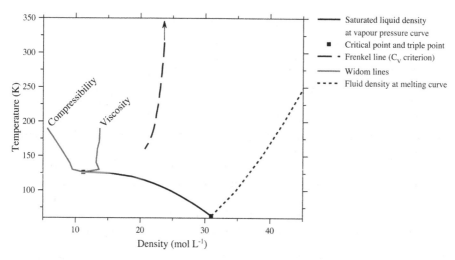

FIGURE B.23 ρ, T phase diagram of fluid N_2. Data for the vapour pressure curve and Widom lines are obtained from the fundamental EOS [4.3] via NIST REFPROP and the ThermoC code [3.25], the fluid density at the melting curve is from ref. [4.5], and the Frenkel line is calculated from the fundamental EOS using the c_V criterion as described earlier. The arrow indicates the expected continuation of the Frenkel line.

References

B.1 J.W. Leachman, R.T. Jacobsen, S.G. Penoncello and E.W. Lemmon, J. Phys. Chem. Ref. Data **38**, 721 (2009).

B.2 H. Hemmes, A. Driessen and R. Griessen, J. Phys. C.: Solid State Phys. **19**, 3571 (1986).

B.3 R.M. Hazen, H.K. Mao, L.W. Finger and P.M. Bell, Appl. Phys. Lett. **37**, 288 (1980).

B.4 B. Olinger, J. Chem. Phys. **77**, 6255 (1982).

B.5 R.J. Hemley, H.-K. Mao, L.W. Finger, A.P. Jephcoat, R.M. Hazen and C.S. Zha, Phys. Rev. B **42**, 6458 (1990).

B.6 L. Bezacier, B. Journaux, J.-P. Perillat, H. Cardon, M. Hanfland and I. Daniel, J. Chem. Phys. **141**, 104505 (2014).

B.7 R.L. Mills, D.H. Liebenberg and J.C. Bronson, J. Chem. Phys. **63**, 4026 (1975).

B.8 D.A. Young et al., Phys. Rev. B **35**, 5353 (1987).

B.9 W. Press. J. Chem. Phys. **56**, 2597 (1972).

B.10 C.A. Swenson, J. Chem. Phys. **23**, 1963 (1955)

B.11 *The new mineralogy of the outer solar system and the high-pressure behaviour of methane*, Ph.D. thesis of Helen E. Maynard-Casely, University of Edinburgh (2009).

B.12 A. Habenschuss, E. Johnson and A.H. Narten, J. Chem. Phys. **74**, 5234 (1981).

B.13 D. Bessières, S.L. Randzio, M.M. Piñeiro, Th. Lafitte and J.-L. Daridon, J. Phys. Chem. B **110**, 5659 (2006).

B.14 C.-J. Chen, M.H. Back and R.A. Back, Can. J. Chem. **53**, 3580 (1975).

B.15 K.D. Litasov, A.F. Goncharov and R.J. Hemley, Earth and Planetary Sci. Lett. **309**, 318 (2011).

B.16 E.H. Abramson, J. Phys.: Conf. Ser. **950**, 042019 (2017).

B.17 V.M. Giordano and F. Datchi, EuroPhys. Lett. **77**, 46002 (2007).

B.18 F. Datchi et al., Phys. Rev. B **94**, 014201 (2016).

B.19 I.A. Richardson and J.W. Leachman, J. Phys. Chem. Ref. Data **43**, 013103 (2014).

B.20 I.A. Richardson and J.W. Leachman, AIP Conf. Proc. **1434**, 1841 (2012).

B.21 V. Diatschenko, C.W. Chu, D.H. Liebenberg, D.A. Young, M. Ross and R.L. Mills, Phys. Rev. B **32**, 381 (1985).

B.22 E. Gregoryanz, A.F. Goncharov, K. Matsuishi, H.-K. Mao and R.J. Hemley, Phys. Rev. Lett. **90**, 175701 (2003).

B.23 K. Trachenko, V.V. Brazhkin and D. Bolmatov, Phys. Rev. E **89**, 032126 (2014).

B.24 R.T. Howie, P. Dalladay-Simpson and E. Gregoryanz, Nat. Mater. **14**, 495 (2015).

B.25 C.-S. Zha, H. Liu, J.S. Tse and R.J. Hemley, Phys. Rev. Lett. **119**, 075302 (2017).

B.26 S.A. Bonev, E. Schwegler, T. Ogitsu and G. Galli, Nature **431**, 669 (2004).

B.27 M.A. Morales, C. Pierleoni, E. Schwegler and D.M. Ceperley, PNAS **107**, 12799 (2010).

B.28 E. Wigner and H. B. Huntington, J. Chem. Phys. **3**, 764 (1935).

B.29 J.M. McMahon, M.A. Morales, C. Pierleoni and D.M. Ceperley, Rev. Mod. Phys. **84**, 1607 (2012).

B.30 W. Wagner, A. Saul and A. Pruβ, J. Phys. Chem. Ref. Data **23**, 515 (1994).

B.31 N. Dubrovinskaia and L. Dubrovinsky, High Press. Res. **23**, 307 (2010).

B.32 A.F. Goncharov et al., Phys. Rev. Lett. **94**, 125508 (2005).

B.33 T. Kawamoto, S. Ochiai and H. Kagi, J. Chem. Phys. **120**, 5867 (2004).

B.34 Y. Katayama et al., Phys. Rev. B **81**, 014109 (2010).

B.35 Th. Strässle et al., Phys. Rev. Lett. **96**, 067801 (2006).

B.36 L.M. Maestro et al., Int. J. Nanotech. **13**, 667 (2016).

B.37 A.S. Zinn, D. Schiferl and M. Nicol, J. Chem. Phys. **87**, 1267 (1987).

B.38 G. Weck et al., Phys. Rev. Lett. **119**, 235701 (2017).

B.39 S. Jiang et al., Nat. Comm. **9**, 2624 (2018).

Appendix C: Some Thermodynamic and Diffraction Derivations

C.1 Thermodynamic Quantities

In this section we outline for reference the definition of the enthalpy, the Gibbs function, and the Helmholtz function, and the conditions for which we refer to them. We follow the treatment given in Roberts and Miller [8.1] but the matter is well described in other standard thermodynamics textbooks also.

C.1.1 Application of the First Law of Thermodynamics

We begin by applying the first law of thermodynamics to a system interacting with its surroundings. We define U as the total internal energy of the system, dQ as the heat flow into the system, and dW as the work done on the system by its surroundings during an infinitesimal change:

$$dU = dQ + dW \tag{C.1}$$

The entropy of the system is S and the entropy of the surroundings is S_0. According to the second law of thermodynamics,

$$dS + dS_0 \geq 0 \tag{C.2}$$

We consider the case where any changes in the surroundings are reversible, in which case we may also write:

$$dS_0 = -\frac{dQ}{T}$$
$$dS - \frac{dQ}{T} \geq 0 \tag{C.3}$$

And substitute with equation D.1 to obtain:

$$dU - TdS \leq dW \tag{C.4}$$

C.1.2 Adiabatic Changes; Enthalpy

In an adiabatic change the entropy of the system cannot decrease: If $dQ = 0$ then we obtain $dU = dW$ and equation C.4 leads to $dS \geq 0$. In addition, the change in the product (PV) must come at the expense of the internal energy U of the system:

$$dW = dU = -d(PV)$$
$$d(U + PV) = 0$$

We therefore introduce the quantity of enthalpy H which remains constant during an adiabatic change, as we proved in Section 6.3.

$$H = U + PV \tag{C.5}$$

For an ideal gas, an adiabatic change is therefore also isothermal.

Let us conclude by examining the condition of the system being in equilibrium. Any infinitesimal change to take the system out of equilibrium must be reversible, because an irreversible change must always lead towards equilibrium. In this case, for such a change taking the system out of equilibrium the inequality in the expression above $dS \geq 0$ becomes an equality and the change must therefore satisfy the condition $dS = 0$. Therefore the entropy of a thermally isolated system is at a maximum when it is in equilibrium.

C.1.3 Isothermal Changes; Helmholtz Function

If the temperature is constant then equation C.4 may be rewritten as:

$$d(U - TS) \leq dW \tag{C.6}$$

It is therefore appropriate to introduce the Helmholtz function F:

$$F = U - TS$$
$$dF \leq dW \tag{C.7}$$

The physical interpretation of the Helmholtz function is that dF represents the minimum amount of work that must be done on the system in the infinitesimal isothermal change. This minimum corresponds to the case where the change in the system (not just the surroundings) is reversible. Furthermore, in a change at constant volume as well as constant temperature, $dW = 0$ and therefore $dF \leq 0$; the Helmholtz function must always decrease or stay constant in such changes.

We can also examine the Helmholtz function at isothermal equilibrium. Any infinitesimal change away from the equilibrium point must by necessity be reversible, because an irreversible change must always bring the system closer to equilibrium. A reversible change corresponds to the equality in equation C.7 so we may write for an isothermal change away from equilibrium:

$$dF = dW \qquad \text{(C.8)}$$

That is to say, that the Helmholtz function's value can only decrease if the system does an exactly equivalent amount of work. Furthermore, if the change is isochoric then the system cannot do any work and the Helmholtz function cannot change. We would write $dF = 0$ for such a change. F has an extremal value at equilibrium and we can conclude from the inequality in equation C.7 that it is a minimum rather than a maximum.

C.1.4 Isobaric and Isothermal Changes; Gibbs Function

In an infinitesimal change at constant pressure, the work done on the system is $dW = -PdV$ and equation C.4 becomes:

$$dU - TdS + PdV \leq 0$$

Therefore, in a change at constant temperature as well as constant pressure:

$$d(U - TS + PV) \leq 0$$

We can now define the Gibbs function G. In a change at constant temperature and constant pressure the Gibbs function of the system is reduced or remains constant (in the case of a reversible change).

$$G = U - TS + PV$$
$$dG \leq 0 \qquad \text{(C.9)}$$

In this case also, an infinitesimally small change moving the system out of equilibrium must be, by definition, reversible so $dG = 0$ in this case. Thus the Gibbs function has an extremal value at equilibrium and we can conclude from inspection of the inequality in equation C.9 that this must be a minimum rather than a maximum.

C.1.5 Constraints on the P, V, T EOS of the Ideal Fluid and the Condensing Fluid (Brown's Conditions)

Here we outline the set of conditions proposed by Brown [3.27] that should be fulfilled by all fluids. We will need to refer to certain limiting points on the phase diagram frequently so will label these using the nomenclature given in Table C.1.

TABLE C.1   Identifying symbols for specific P, T conditions referred to in the text

Identifier	Conditions
Π_1	Low temperature limit
Π_2	High pressure limit
Π_3	High temperature limit
Π_4	Low pressure limit
I	High temperature and low pressure limit
J	Low temperature and high pressure limit
K_L (liquid) and K_G (gas)	Low temperature and low pressure limit
L	High temperature and high pressure limit
O	Critical point

To begin we must implement for mathematical convenience logarithmic forms of the reduced thermodynamic variables as defined in equation 2.19. We implement the compression factor C and reduced compression factor C_R, also with a logarithmic representation. The compression factor at the critical point is denoted by C_{CP}.

$$\tilde{P} = \ln P_R \qquad\qquad\qquad (C.10)$$

$$\tilde{V} = \ln V_R$$

$$\tilde{T} = \ln T_R$$

$$C = \frac{PV}{RT}$$

$$C_R = \frac{P_R V_R}{T_R}$$

$$C_{CP} = \frac{P_C V_C}{RT_C}$$

$$\tilde{C} = \ln C_R$$

Hence the logarithmically represented variables all equal zero at the critical point. We are going to refer to two specific samples: The ideal fluid and the condensing fluid. The ideal fluid is distinguished from the ideal gas by the axiom of the extension of mass. This is the condition that, even at Π_2, $V_R > 0$ and therefore $\tilde{V} > -\infty$. The ideal fluid is one in which there are repulsive interactions between particles. The condensing fluid is one in which there are also attractive interactions between particles, and which therefore *condenses*. We need to define in useful form some limits on the behaviour of $\tilde{C}$ at certain conditions for all fluids, and then for condensing fluids only. By definition, the ideal fluid does not have a critical point. Thus, whilst for the condensing fluid P_R, V_R, T_R represent the parameters at the critical point, for the ideal fluid they represent an arbitrary P, T point that can be anywhere except the limits described in Table C.1.

For all fluids:

Condition 1
The fluid behaviour must tend towards ideal gas behaviour $(C \to 1)$ at Π_4, but this condition does not need to be satisfied at Π_3.

Condition 2
At J, $\tilde{C} \to \infty$ since $V_R \neq 0$ as described above.

Condition 3
At I:

$$\left(\frac{\partial \tilde{C}}{\partial \tilde{T}}\right)_P = \left(\frac{\partial \tilde{C}}{\partial \tilde{T}}\right)_V = \left(\frac{\partial \tilde{C}}{\partial \tilde{P}}\right)_T = \left(\frac{\partial \tilde{C}}{\partial \tilde{P}}\right)_V \to 0$$

This is because, at I, we must recover ideal gas behaviour in which case $\tilde{C}$ is constant. It is more stringent than condition 1 which is satisfied all along Π_4. This is because I is the absolute low density limit, in which case we do not expect any change in P,T conditions to affect C.

Condition 4
At Π_1 and Π_2:

$$\left(\frac{\partial \tilde{C}}{\partial \tilde{T}}\right)_P = \left(\frac{\partial \tilde{C}}{\partial \tilde{T}}\right)_V \to -1$$

Dealing with the first derivative:

$$\left(\frac{\partial \tilde{C}}{\partial \tilde{T}}\right)_P = \left(\frac{\partial \tilde{C}}{\partial C_R}\right)_P \left(\frac{\partial C_R}{\partial T_R}\right)_P \left(\frac{\partial T_R}{\partial \tilde{T}}\right)_P = \frac{T_R}{V_R}\left(\frac{\partial V_R}{\partial T_R}\right)_P - 1$$

At Π_1 this condition is clearly satisfied. At Π_2:
$\left(\frac{\partial V_R}{\partial T_R}\right)_P \to 0$ but $V_R \neq 0$ due to the axiom of extension of mass allowing the condition to be satisfied.

Dealing with the second derivative:

$$\left(\frac{\partial \tilde{C}}{\partial \tilde{T}}\right)_V = \left(\frac{\partial \tilde{C}}{\partial C_R}\right)_V \left(\frac{\partial C_R}{\partial T_R}\right)_V \left(\frac{\partial T_R}{\partial \tilde{T}}\right)_V = \frac{T_R}{P_R}\left(\frac{\partial P_R}{\partial T_R}\right)_V - 1$$

As a result this is clearly satisfied at Π_1 and Π_2.

Condition 5
At Π_2:

$$\left(\frac{\partial \tilde{C}}{\partial \tilde{P}}\right)_T \to 1$$

This can be proved as follows:

$$\left(\frac{\partial \tilde{C}}{\partial \tilde{P}}\right)_T = \left(\frac{\partial \tilde{C}}{\partial C_R}\right)_T \left(\frac{\partial C_R}{\partial P_R}\right)_T \left(\frac{\partial P_R}{\partial \tilde{P}}\right)_T = 1 + \frac{P_R}{V_R}\left(\frac{\partial V_R}{\partial P_R}\right)_T$$

Due to the axiom of the extension of mass P_R / V_R remains finite at Π_2 allowing condition 3 to be satisfied due to:

$$\left(\frac{\partial V_R}{\partial P_R}\right)_T \to 0 \text{ at } \Pi_2.$$

Condition 6
At Π_1:

$$\left(\frac{\partial \tilde{C}}{\partial \tilde{P}}\right)_T \to 1 \text{ in a monotonic increase.}$$

Considering the expression obtained in our proof of condition 5 above, we can see that the limiting value at Π_1 is 1. In addition, since:

$$\left(\frac{\partial V_R}{\partial P_R}\right)_T < 0$$

The limiting value of 1 must be approached by an increase from a lower value.

Condition 7
For the condensing fluid only, $C_R \to 1$ and $\tilde{C} \to 0$ at O.

C.2 Fourier Transform Treatment of Diffraction

It is relatively straightforward to perform a general treatment of diffraction from a sample which shows that the diffraction pattern is the Fourier transform of the distribution of atoms in the fluid in real space[*]. Once this feature is understood, most features of diffraction experiments become self-evident. As a reference point, let us recall the relationship between a function $f(v)$ and its Fourier transform $F(w)$:

$$F(w) = \frac{1}{(2\pi)^{3/2}} \int f(v) e^{iv.w} dv$$

$$f(v) = \frac{1}{(2\pi)^{3/2}} \int F(w) e^{-iv.w} dw$$

(C.11)

---

[*] This is subject to the assumption that we can precisely subtract the background from the diffraction pattern and account for other experimental constraints.

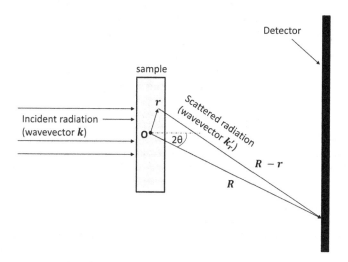

FIGURE C.1 Experimental geometry for a diffraction experiment.

Here, we can of course choose how the $1/(2\pi)^{3/2}$ factor is split between the two transforms and on which exponent we place the minus sign. Let us (adapting the approach of ref. [5.12]) consider the sample as a function $\rho(r)$ representing the amplitude of scattering of the incident radiation at each location in the sample. So $\rho(r)d\tau$ is the amplitude of scattering from the part of the sample within the volume element $d\tau$ located at r.

Let us consider (Figure C.1) the amplitude dE of the diffracted beam at a point R, from the volume element $d\tau$. We can write:

$$dE = E_0 e^{i(k.r-\omega t)} \times \rho(r)d\tau \times \frac{e^{ik|R-r|}}{|R-r|} \tag{C.12}$$

Recalling that $k = k_r'$ due to the elastic scattering and that $(R-r)$ and k_r' point in the same direction, we can manipulate the phase factor in the exponential as follows:

$$k|R-r| = k_r'.(R-r) = k_r'.R - k_r'.r \approx kR - k_r'.r$$

Making these substitutions we obtain:

$$dE = \rho(r)d\tau e^{-iQ.r} \tag{C.13}$$

Where we have ignored the factors that are the same for all locations in the sample and defined the scattering vector Q:

$$Q = k' - k$$

Where now we use the notation k'_r to denote the wavevector of the beam scattered from location r. The expression in equation (C.13) can be integrated to obtain the total diffracted amplitude for a given R (i.e., for a given Q).

$$E(Q) = \int \rho(r)e^{-iQ \cdot r}d\tau \tag{C.14}$$

Comparing the expression above to equation (C.11) we can see that $E(Q)$, the function representing the diffracted amplitude, is simply the Fourier transform of the real space lattice as represented by $\rho(r)$. The structure factor[†] is the scattered *intensity* per unit atom $F(Q)$ and is obtained from $E(Q)$:

$$F(Q) = \frac{1}{N}|E(Q)|^2 \tag{C.15}$$

Where there are N atoms in the sample. Note that this definition of the structure factor is not quite consistent with equation 4.4.

$\rho(r)$ depends on what probe we are diffracting from the sample. X-rays are scattered entirely by the electrons in the sample, so for an X-ray diffraction experiment $\rho(r)$ would be the distribution of electronic charge throughout the sample. This can be approximated by a δ-function at the location of each nucleus multiplied by the atomic form factor which decreases for increasing Q so is written $f(Q)$. In a neutron diffraction experiment on the other hand, $\rho(r)$ would be a δ-function at the location of each nucleus multiplied by a constant, the scattering length b_j. In both cases we label the positions of the atoms with R_j. This qualitative difference is due to X-rays being scattered only from the electrons in the sample (via the electromagnetic force) whilst neutrons are scattered only from the nucleus, due to the strong nuclear force.

$$\rho_{neutron}(r) = \sum_{j=1}^{N} b_j \delta(r - R_j)$$
$$\rho_{X-ray}(r) = \sum_{j=1}^{N} f(Q) \delta(r - R_j) \tag{C.16}$$

Recalling the properties of the δ-function we obtain the following expressions for $E(Q)$ for neutron and X-ray diffraction:

$$E_{neutron}(Q) = \sum_{j=1}^{N} b_j e^{-iQ \cdot R_j}$$
$$E_{X-ray}(Q) = \sum_{j=1}^{N} f(Q) e^{-iQ \cdot R_j} \tag{C.17}$$

[†] Note that the term "structure factor" has a different meaning here to its meaning when considering diffraction from solids.

Appendix D: The Diamond Anvil Cell (DAC)

In recent years the diamond anvil cell (DAC) has been extensively utilized for research on dense supercritical fluids. This has followed from the proposal of the Frenkel line causing increased interest in the supercritical fluid state at liquid-like or even solid-like densities and developments at the new generation of synchrotron light sources. In the last decade or so the X-ray spot sizes achieved at these facilities have decreased drastically [4.19] allowing us to probe ever-smaller samples inside the DAC at ever-more extreme pressures. The development of in situ laser heating instruments on synchrotron beamlines has allowed the generation of high temperatures of up to ca. 5000 K. At 5000 K the DAC is the only device that can generate sufficient pressure to reach melting curves of typical materials.

Additionally, in recent years there has been a crossover in experimental methods between the X-ray and neutron scattering communities. Traditionally, high-pressure X-ray diffraction was performed on samples in the DAC and high-pressure neutron diffraction was performed on bigger samples contained in large volume devices such as the Paris-Edinburgh cell. However, in recent years, due to improvements in the technology to focus neutrons, it has become possible to perform neutron diffraction in the DAC [D.1] (albeit in DACs with unusually large diamonds!). Instrumentation used to improve the signal-to-noise ratio in neutron scattering experiments on fluids in large volume cells (Soller slits) have been adapted for X-ray diffraction in the DAC.

In this appendix we will therefore review the basic mechanics of the DAC and its capabilities and limitations.

D.1 Design of the DAC

The diamond anvil cell (Figure D.1 (a)) consists of two opposing diamonds with the culets precisely aligned, mounted on an arrangement such as a piston and cylinder to maintain the alignment. Force to generate high pressure between the diamond tips is applied along the axes of the diamonds as shown. This force is generated using relatively simple methods: A gas membrane, hydraulic ram, lever arm, or simply four socket cap bolts.

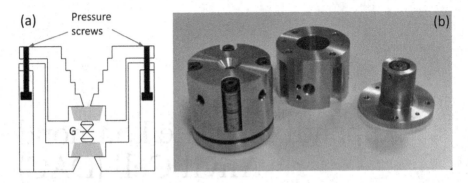

FIGURE D.1   (a) Schematic diagram of the DAC. The label G indicates the gasket between the diamonds. (b) Assembled and disassembled DACs.

Figure D.1 (b) shows the DAC as assembled. Simply turning the screws can generate pressures of over 100 GPa (1 million atmospheres).

All DAC research nowadays takes place with a metal gasket between the diamonds as shown in Figure D.1 (a). A hole in the gasket centered as well as possible over the diamond culets contains the sample, a pressure calibrant, and (in some cases) a pressure transmitting medium. The smaller the diamond culet diameter, the higher the pressure that may be attained in the DAC before the diamonds break. Figure D.2 shows the maximum

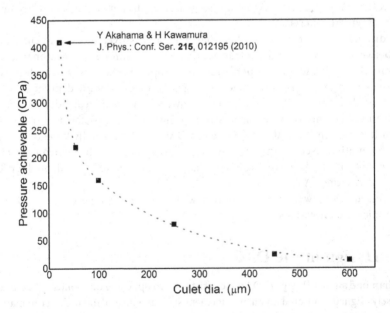

FIGURE D.2   Maximum pressure attainable in the DAC as a function of the culet diameter [D.3]. Reprinted from J.E. Proctor and D. Massey, Rev. Sci. Instrum. **89**, 105109 (2018), with the permission of AIP Publishing.

attainable pressure as a function of the culet diameter. All but one of the datapoints correspond to diamonds that the author has personally broken over the years. In recent years significantly higher static pressures have been attained using double-stage DACs [D.2], but this technology is in its infancy—it is likely that years of development work will be required before these devices are of any practical use for the study of fluids. We shall therefore not review them here.

D.2 Loading of Fluid and Fluid Mixture Samples into the DAC

This seemingly mundane topic warrants coverage here as some experimental methods to load samples into the DAC are intrinsically vulnerable to contamination problems, as discussed in the literature recently [D.4]. We will discuss pure fluids first, then move on to the additional problems posed by fluid mixtures.

D.2.1 Pure Fluids

Let us begin with samples which are solids at ambient conditions, but which will melt when heated in the DAC. If high temperature is to be generated using resistive heating (typically up to 1000 K) the DAC sample chamber can simply be filled with the sample at ambient conditions under a microscope with a fine needle and good hand-eye co-ordination. If high temperature is to be generated using laser heating (typically preferred for 1000 K and higher) it is necessary to insulate the sample from the diamonds and gasket. In a laser heating experiment the heating lasers are tightly focussed on the sample and heating of the diamonds is to be avoided if at all possible. This is essential to avoid diamond breakage and also ensures the sample reaches the highest temperature possible. This insulation is achieved by placing layers of soft, unreactive, transparent, and thermally insulating material all around the sample; which must be significantly smaller than the sample chamber diameter.

A soft material is preferred as it will support less shear stress, leading to more hydrostatic conditions and less variation in pressure throughout the sample chamber. Salts such as NaCl usually work well. Figure D.3 (a) shows a Vanadium sample loaded into the DAC sample chamber ready for a laser heating study of the melting curve at the European Synchrotron Radiation Facility [D.5]. What looks like empty space around the sample is in fact NaCl, which turns transparent under very modest compression.

Samples which are liquid at ambient conditions can be loaded into the DAC sample chamber simply by placing a droplet in the chamber with a fine needle before closing the DAC, but samples which are gaseous at ambient conditions provide perhaps the greatest challenge. The compressibility of gases is such that if loaded at ambient conditions or even gas cylinder pressure of ca. 25 MPa the sample chamber will shrink to an impracticably small size before liquid-like density is reached, let alone solidification. It is thus necessary to increase the density of the gas significantly before loading, using either low temperature or high pressure.

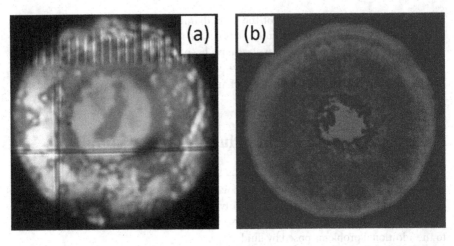

FIGURE D.3 (a) Vanadium sample surrounded by NaCl in the DAC sample chamber in preparation for laser heating experiment at the European Synchrotron Radiation Facility. (b) Solid ethane and ruby microcrystal in the DAC sample chamber. Panel (a) is author's own photograph and panel (b) is courtesy of the University of Salford Maker Space.

In principle, any sample which is gaseous at ambient conditions can be liquefied in the laboratory by cooling using liquid He. In practice, liquid N_2 is used to liquefy gases that can be liquefied by cooling to ca. 80 K or above, and gases that would require a lower temperature to liquefy are densified using high pressure instead. Liquefaction of gases for DAC loading using liquid N_2 is usually performed using apparatus similar to that shown in Figure D.4. Here, the entire DAC is held inside the container as shown

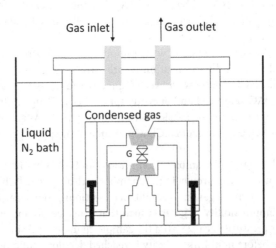

FIGURE D.4 Cryogenic loading apparatus for the DAC. The gas is condensed into the liquid state (labelled "condensed gas") by the cooling provided from the surrounding liquid N_2 bath.

and cooled to liquid N_2 temperature. The container holding the DAC is flushed out with the gas to be loaded before cooling commences and the DAC is completely immersed in liquefied gas. During cooling the DAC must be open so that the liquid can flow into the sample chamber. This is usually achieved by placing a spring between the piston and cylinder. Then, pressure is applied to close the DAC whilst it is immersed in the liquid. The pressure can be applied using a He-driven gas membrane or simply by opening the container and reaching in with long Allen keys to turn the screws.

This DAC loading method normally requires the condensation of a large quantity of the gas being loaded into the DAC. In some cases (for example SiH_4 [D.6], F_2 [D.7]) this may not be possible for safety reasons. For gases such as these an alternative is to cool the DAC in an inert atmosphere with a lower condensation point than the gas being loaded, then inject a tiny quantity of the gas to be loaded (via a nozzle) to condense in and around the sample chamber. The DAC is then closed and placed under pressure before warming back up to 300 K. The risk of contamination with N_2, H_2O, or the gas used as an inert atmosphere are much higher with this method so it should only be attempted when absolutely necessary.

As an alternative to low temperature, high pressure can be utilized to load samples into the DAC that are gaseous at ambient conditions. Generally, this is only used for gases (H_2, He, and Ne) which cannot be liquefied by cooling with liquid N_2. In this method, the (open) DAC is placed inside a pressure chamber which is pumped up to (at least) 15 MPa (1500 bar) pressure with the gas to be loaded into the DAC. Under this pressure the DAC is closed to trap a sufficient quantity of the gas in the sample chamber. A simple mechanical feedthrough to close the DAC by turning the pressure screws via a gearbox is most fail-safe, but closure of the DAC using a gas membrane is also performed. This is difficult since the pressure differential between the gas pressure in the pressure chamber and the gas membrane needs to be kept constant as pressure is increased to 15 - 20 MPa.

D.2.2 Fluid Mixtures

The loading of fluid mixtures into the DAC presents special experimental difficulties. Gas mixtures can be loaded cleanly using high pressure alone as described in the previous section, but cryogenic loading of a gas mixture is potentially problematic if the gases have different condensation temperatures at ambient pressure. Loadings of gas-liquid mixtures are also possible. A drop of the liquid, not filling the sample chamber, can be loaded and the DAC sealed shut. Then the DAC can be opened using a spring under high gas pressure or whilst immersed in the liquefied gas. But these experiments are very difficult. Accurately determining the ratio of different fluids inside the sample chamber is often not possible.

D.3 High Temperatures in the DAC

Samples can be heated in situ under high pressure in the DAC using either resistive heating or laser heating. In summary, resistive heating results in more accurate control and measurement of temperature whilst laser heating allows higher temperatures to be obtained. Here we will review the experimental methods and results obtainable with both methods. Pressure measurement at high temperature is dealt with in Section D.4; this also presents some experimental problems which cannot be ignored.

A general comment, applying to both resistive heating and laser heating experiments, is that maintaining a constant temperature and pressure for extended periods of time is challenging. Data collection for a single P, T datapoint when performing Raman spectroscopy or (elastic) X-ray scattering typically takes ca. 1 minute, but data collection for a single P, T datapoint in an inelastic X-ray scattering experiment and any neutron scattering experiment takes far longer—typically several hours. Thus some experiments are easier than others to perform at high temperature in the DAC.

D.3.1 Resistive Heating Experiments in the DAC

Using resistive heating, high temperatures of up to 1000 K can be routinely achieved in the DAC, and temperatures of up to 1200 K are possible [D.8]. Typically, this is performed using two separate heaters: The first heater is clamped around the entire DAC and can heat it up to about 550 K unaided. A thermocouple is attached somewhere close to the sample chamber, for instance to the diamond seat, using a thermally conductive glue. Ideally the thermocouple reading should be utilized to control the power to the heater in order to ensure a constant temperature. Figure D.5 shows this simple arrangement in operation in the author's lab.

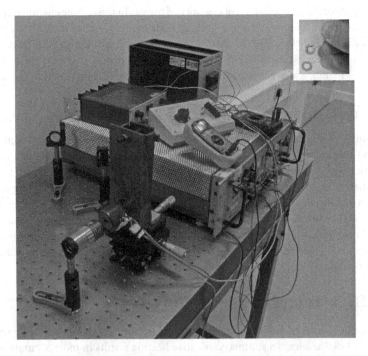

FIGURE D.5 DAC equipped with external resistive heater and temperature controller for Raman spectroscopy at high pressure and high temperature. Inset: Internal heater under construction.

To achieve higher temperature the heater surrounding the DAC is augmented by a second, ring-shaped, heater which sits inside the DAC on top of the piston. The diamonds and gasket sit within the ring and thermally conductive glue is used to ensure the heat is transmitted as effectively as possible to the sample chamber. Due to the smaller area heated and consequent potential for temperature gradients within the cell, the potential for poorly-quantified error in temperature measurement is higher. In addition, due to the smaller volume being heated, a change in current through the heater can take effect very quickly so feedback to ensure a constant temperature is harder to implement. The second heater is typically a single-use item. The inset to Figure D.5 shows one of these heaters under construction. The nichrome or tungsten wire is wound around a ceramic ring before being enclosed in a thermally conductive cement.

As an alternative to temperature measurement using a thermocouple, in the specific case of a Raman spectroscopy experiment it is possible to measure temperature from the intensity ratio of the Stokes and anti-Stokes branches of the same Raman peak from the sample (or some other material placed in the sample chamber with the sample), provided that the energy E of the peak is sufficiently low that the anti-Stokes branch can be observed with significant intensity at ambient temperature (i.e., $E \approx k_B T$).

Diamond starts to oxidize above ca. 650 K. To conduct experiments above this temperature it is necessary to prevent this by blowing an inert gas over the DAC, or by enclosing the DAC in a vacuum. The latter option also assists in enabling higher temperatures by removing one means for conduction of heat away from the DAC [D.8].

D.3.2 Laser Heating in the DAC

Alternately, one can heat samples in the DAC using radiation from a high power infrared laser(s). Typically, this method is used to heat to temperatures of ca. 2000 K upwards. Subjecting the diamonds to these temperatures would instantly destroy them simply due to thermal decomposition or reaction with the contents of the sample chamber, regardless of how effectively the DAC is contained in a vacuum or inert atmosphere; the author knows this from personal experience. Laser heating experiments in the DAC therefore involve heating the sample inside the pressure chamber whilst keeping the diamonds as close as possible to 300 K.

Lasers for high-temperature experiments in the DAC are selected for a wavelength at which the sample absorbs light as strongly as possible. So, for instance, a Nd:YAG (Neodymium-doped Yttrium Aluminium Garnet) laser at 1.064 µm would be a good choice for studying the melting curves of transition metals, or a CO_2 laser at 10.6 µm for the melting curves of wide-bandgap materials such as silicates [D.9]. Alternatively, an absorber (usually a ring-shaped section of a transition metal foil) can be placed in the sample chamber to absorb the laser light then transmit heat to a weakly-absorbing sample. This was used, for instance, in the laser heating study of melting and high-temperature ionization of H_2O cited in Appendix B [B.34]. The use of beam absorbers can be problematic for an X-ray diffraction study as they are made from high-Z materials that can give rise to very intense Bragg peaks, interfering with the weak diffraction peaks from the fluid sample even if only a tiny proportion of the incident X-ray beam is hitting the absorber rather than the sample within.

There are essentially two options to measure the sample temperature in situ during a laser heating experiment: Fitting to the blackbody radiation emitted by the sample using Planck's law, or measurement of the Stokes/anti-Stokes intensity ratio of a selected Raman peak as discussed earlier in relation to resistive heating. The sample properties in a Raman spectroscopy or X-ray scattering experiment are probed using a beam of diameter ca. 1 μm. The heating laser beam is typically focussed to a somewhat wider spot size of ca. 10 μm. Nevertheless, temperature gradients in the sample are potentially large and hard to quantify.

D.4 Pressure Measurement in the DAC

The force applied to the diamonds to generate pressure in a DAC experiment can be measured accurately if it is applied using a hydraulic ram. Sometimes this is used for diagnostic purposes. However, the gasket away from the diamond culets and sample chamber is typically 10x the thickness at the sample chamber and the gasket covers a large portion of the facets of the diamonds as well as the culets. This is essential to ensure stability at high pressure. In this case, the force applied is not uniform over a known area and it is not possible to directly calculate the pressure using $P = F / A$. Pressure must instead be measured by making a photoluminescence or diffraction measurement on a pressure marker placed in the sample chamber with the sample. The photoluminescence method is most relevant to this text.

In experiments at 300 K the most commonly utilized photoluminescence pressure markers are ruby and Sm:YAG (Samarium-doped YAG). Both provide accuracy just about adequate for the collection of closely spaced data points at pressures up to ca. 2000 MPa that are usually required for the study of fluids. The photoluminescence peaks utilized from ruby originate from electronic transitions in the Chromium ions with which the Al_2O_3 is doped to form ruby, at ca. 695 nm. In fact, these are the transitions that give ruby its characteristic red colour. Figure D.6 (a) shows example spectra of ruby at ambient conditions and at ca. 1000 MPa pressure. The most accurate pressure measurement is obtained under completely hydrostatic conditions resulting from the entire sample chamber being filled with fluid. In this case (see Figure D.6 (a)) the Lorentzian fits to the overlapping peaks result in an error in the measured pressure of ca. 1 MPa using the known calibration [D.10]. In reality, the spread of data about the line of best fit in the experiment usually indicates that this does not account for all sources of error.

The use of Sm:YAG photoluminescence to measure pressure may be preferred at 300 K if the photoluminescence peaks from ruby overlap with Raman peaks from the sample (the photoluminescence peaks from both ruby and Sm:YAG are vastly more intense than the most intense Raman peaks (H_2, CH_4 etc.)). However, Sm:YAG's principal utility is in pressure measurement at high temperature during resistive heating experiments. The photoluminescence peaks from both ruby and Sm:YAG significantly broaden and weaken upon temperature increase. The ruby broadening is far worse and the problem is compounded by the peaks shifting upon heating (in addition to the shift due to applied pressure). Figure D.6 (b) shows the photoluminescence peaks from Sm:YAG that are used to measure pressure, at 300 K and at 536 K, both at 1 bar pressure. The weakening

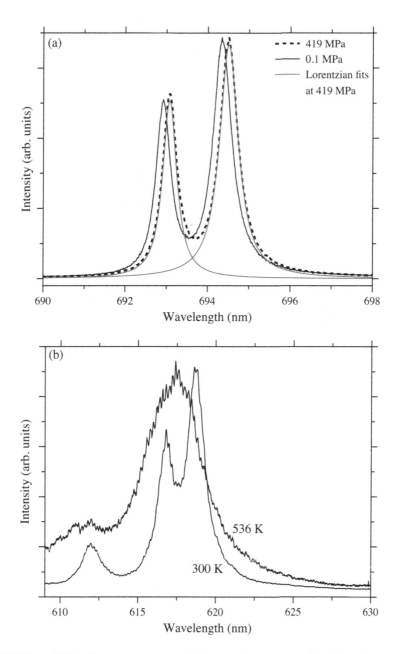

FIGURE D.6 (a) Photoluminescence spectra of ruby at ambient pressure of 0.1 MPa and at 419 MPa, at 300 K. In the latter case the Lorentzian fits performed to measure pressure are illustrated. (b) Photoluminescence spectra of Sm:YAG at 0.1 MPa, 300 K and 536 K. Spectra in panels (a) and (b) are normalized to the same intensity. In reality in panel (b) the spectrum at 536 K has significantly lower intensity than at 300 K.

and broadening of the peaks that ultimately limits what can be achieved in high-temperature DAC experiments in the 0–2000 MPa range is illustrated.

In laser heating experiments, the most common approach taken is to measure pressure from a part of the sample chamber which is expected to remain at 300 K. However, the variation in pressure throughout the sample chamber in a laser heating experiment can easily exceed 1000 MPa [B.34].

References

D.1 R. Boehler, J.J. Molaison and B. Haberl, Rev. Sci. Instrum. **88**, 083905 (2017).

D.2 N. Dubrovinskaia et al., Sci. Adv. **2**, e1600341 (2016).

D.3 J.E. Proctor and D. Massey, Rev. Sci. Instrum. **89**, 105109 (2018).

D.4 R. Turnbull et al., Phys. Rev. Lett. **121**, 195702 (2018).

D.5 D. Errandonea et al., Phys. Rev. B **100**, 094111 (2019).

D.6 M. Hanfland, J.E. Proctor, C.L. Guillaume, O. Degtyareva and E. Gregoryanz, Phys. Rev. Lett. **106**, 095503 (2011).

D.7 D. Schiferl and S. Kinkead, J. Chem. Phys. **87**, 3016 (1987).

D.8 Z. Jenei, H. Cynn, K. Visbeck and W.J. Evans, Rev. Sci. Instrum. **84**, 095114 (2013).

D.9 D. Smith et al., Rev. Sci. Instrum. **89**, 083901 (2018).

D.10 A. Dewaele, M. Torrent, P. Loubeyre and M. Mezouar, Phys. Rev. B **78**, 104102 (2008).

Appendix E: Code for Selected Computational Problems

In this appendix computer code is given for selected computational problems referred to in the main text, and the coding methodology is explained. The code has been written in Octave (an open-source version of Matlab) but should be easily adaptable to other languages. It is available on request to the author.

E.1 Boiling Transition in the van der Waals Fluid

The code described here is for calculating the boiling transition P,T conditions predicted from a cubic EOS. The van der Waals EOS is used for simplicity but, with the caveats mentioned later, the code should be adaptable to other cubic EOS. The boiling or condensation phase transition takes place at the P,T conditions for which it results in no change to the Gibbs function of the fluid sample. We can begin by recalling that, for a transition from state A to state B along an isotherm [1.5][2.8]:

$$\Delta G = \int_{A}^{B} V dP \tag{E.1}$$

We will work with the example of the isotherm at 180 K of a van der Waals fluid with $T_c = 200$ K and $P_c = 5$ MPa (Figure E.1). Setting these parameters results in the determination of a, b, and V_c according to the usual van der Waals' formulae (equations 2.16–2.18). We can see by inspection of equation E.1 that $\Delta G = 0$ for a transition in which the two areas separating the dotted line from the isotherm are equal [1.5].

Therefore, we may write the condition as:

$$\Delta G = \int_{V_l}^{V_g} (P - P_b) dV = 0 \tag{E.2}$$

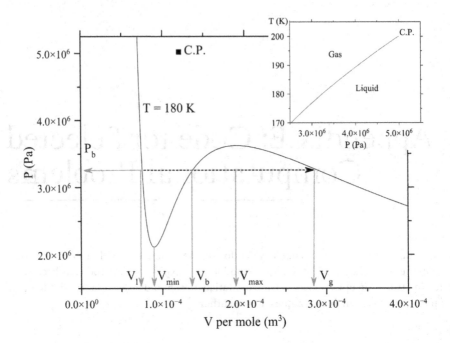

FIGURE E.1　Van der Waals isotherm at 180 K for fluid with $T_c = 200$ K and $P_c = 5$ MPa (black square shows the critical point). Inset: Vapour pressure curve predicted from 170 K to T_c using the van der Waals EOS.

Where $P(V)$ is given by equation 2.14 but we do not know the limits for the integral, or P_b—in fact calculating P_b is our objective. What we can do is to estimate P_b, use this estimated value to obtain the required limits for the integral in equation E.2, then revise our estimate of P_b in the direction required to reduce the magnitude of ΔG.

E.1.1 Estimate of P_b

We may make an initial estimate of P_b by differentiating equation 2.14 (reproduced below as equation E.3) to obtain the extrema V_{max} and V_{min}.

$$P = \frac{RT}{V-b} - \frac{a}{V^2} \tag{E.3}$$

$$\left(\frac{\partial P}{\partial V} \right)_T = -\frac{RT}{(V-b)^2} + \frac{2a}{V^3} = 0 \tag{E.4}$$

Equation E.4 is cubic in V, with three real roots for subcritical temperatures. The smallest root is a mathematical artefact and can be discarded, whilst the other two roots correspond to V_{max} and V_{min} in Figure E.1. In procedures where cubic EOS are solved to obtain three roots, one of which is discarded, it is in general necessary to check which root is to be discarded. For the van der Waals EOS the roots always fall in a specific order but this is the exception rather than the rule. We then make our estimate by inspection of the isotherm in Figure E.1: $V_b \approx \frac{1}{2}(V_{max} + V_{min})$ and use equation E.3 to obtain P_b.

E.1.2 Evaluation of ΔG

Substituting the value of P_b into equation E.3 and solving for V, we obtain three real roots for a subcritical isotherm: V_l, V_b, and V_g. If we wished to adapt this routine for a different cubic EOS we would again need to include additional checks to ensure we use the correct roots. Assuming the correct roots have been identified, we now have the required limits to evaluate equation E.2 and obtain ΔG:

$$\Delta G = \int_{V_l}^{V_g} \left(\frac{RT}{V-b} - \frac{a}{V^2} \right) dV - P_b \left(V_g - V_l \right) = RT \ \ln \left(\frac{V_g - b}{V_l - b} \right) + a \left(\frac{1}{V_g} - \frac{1}{V_l} \right) - P_b \left(V_g - V_l \right)$$

By inspection of Figure E.1 we can see that if $\Delta G > 0$ then our estimate of P_b should be revised upwards, and vice versa. Then equation E.3 can be solved to obtain new values of V_l and V_g and a revised value of ΔG is obtained. In the code presented, a simple finite difference method is used to obtain successive values of P_b and ΔG. This achieves convergence at adequate speed, though users who wish to perform very large numbers of calculations may wish to replace it with a routine that converges faster. For the parameters $T_c = 200$ K and $P_c = 5$ MPa the vapour pressure curve can be predicted as shown in the inset to Figure E.1. The output values from each execution of the routine are printed to the screen using the disp function.

E.1.3 Octave Code for van der Waals' Boiling Transition

```
retval = 0;
Tc = 200; #Critical T in K
Pc = 5E6; #Critical P in Pa
R = 8.31451; #Ideal gas constant
T = 180; #T at which boiling pressure is to be calculated.
Vrange = [0, 0, 0]; #V range in which no stable state exists
Vrange2 = [0, 0, 0]; #For storing solutions to cubic eqns
Vmax = 0; #gas volume after boiling
Vmin = 1E6; #liquid volume before boiling.
#Therefore routine cannot deal with boiling at V ~ 1E6
count_var = 0; #Variable for general use
Gibbs_prev = 0; #For storing previous Gibbs calculation
Increment = 0.01; #Can be decreased to achieve convergence
Vb = 0; #Current best guess at boiling V,P etc
Pb = 0;
```

```
Vg = 0;
Vl = 1E6;

I = 0; #Integral that should be zero at the boiling transition
Vgt = 0; #Variables for passing arguments to Gibbs function
Vlt = 0;
Pbt = 0;

if T>Tc
  disp('There is no boiling transition since T > Tc');
end

#Function for calculating change in
#Gibbs function ready for later

function retval = Gibbs (Vgt, Vlt, Pbt, R, T, a, b)
  retval = (R*T*log((Vgt-b)/(Vlt-b))) + (a*((1/Vgt)-(1/Vlt)))…
  … - (Pbt*(Vgt-Vlt));
endfunction

#Now V der W constants a and b are calculated
#cf equations 2.16 and 2.17

Vc = 3*R*Tc/(8*Pc); #cf equation 2.18
a = 9*Vc*R*Tc/8;
b = Vc/3;

#Derive max and min volumes on V der W curve

c = [R*T, -2*a, 4*a*b, -2*a*(b^2)];
Vrange = roots(c);
#Put max volume into the correct variable.
#Without knowledge of the internal working of roots function,
#there is no way to know which is max and min.
#It is best to check.

for n = 1:3
  if Vrange(n) > Vmax
    Vmax = Vrange(n);
  end

  if Vrange(n) < Vmin
    count_var = n;
    Vmin = Vrange(n);
  end
end

Vrange(count_var) = 1E6;
#Discard Vrange(count_var) as this is spurious
Vmin = 1E6;
```

```
for n = 1:3
  if Vrange(n) < Vmin
    Vmin = Vrange(n);
  end
end

#Initial guess at Vb, Pb
Vb = (Vmax + Vmin)/2;
Pb = ((R*T)/(Vb-b)) - (a/(Vb^2));

for q = 1:100
  #Find Vg, Vl, then perform Gibbs function using these limits
  c = [Pb, -((Pb*b)+(R*T)), a, -a*b];
  Vrange2 = roots(c);

  for n = 1:3
    if Vg < Vrange2(n)
      Vg = Vrange2(n);
    endif
    if Vl > Vrange2(n)
      Vl = Vrange2(n);
    endif

  endfor

  Gibbs_prev = I;

  I = Gibbs (Vg, Vl, Pb, R, T, a, b);

  if I!=0 #Needed otherwise the next line is division by zero
  #in the first iteration, and if convergence has been achieved
    if Gibbs_prev/I < 0 #This is true if we have overshot
    #convergence i.e. if the variable I has changed sign
      Increment = Increment/10;
    endif
  endif

  if I > 0
    Pb = Pb*(1+Increment);
  else
    Pb = Pb*(1-Increment);
  endif

endfor

disp('T =');
disp(T);
disp('Pb =');
disp(Pb);
disp('Gibbs ='); #Displays change in Gibbs func. at convergence
disp(I);
```

E.2 Prediction of Fluid Heat Capacity

In Section 5.1.2 we outlined how the experimentally observed heat capacities of a range of fluids can be modelled by accounting for the ability of the fluid to support shear waves with frequency ranging from the Frenkel frequency to the Debye frequency. Results calculated by the author were presented for liquid Hg (Figure 5.5) and fluid Ar (Figure 5.6). Here we present the code used for the liquid Hg heat capacity calculation. It is easily adaptable for other fluids by substituting certain constants [the Debye period τ_D (tau _ d), infinite frequency shear modulus G_∞ (g _ inf), and thermal expansion coefficient β (beta)]. In addition, the viscosity data for the fluid in question needs to be substituted for the Hg data. This can be achieved by substituting a fit to the data equivalent to that in equation 5.13, or by loading a file containing output from the fundamental EOS if one is available for the fluid. To produce the Ar fit (Figure 5.6), a file of viscosity data produced by the fundamental EOS was used to populate the sampleparam(i,2) column. Alternatively, one could fit using the Vogel-Fulcher-Tammann law as performed in the heat capacity calculations in ref. [1.15].

The Debye function (equation 5.8 and Figure 5.3) is implemented as a separate function debye(x) (appended to the end of the code as shown), which in turn calls on the debye _ integrand(z) function to provide the integrand. The numerical integration is performed using the quadgk function built into Octave. Substitution of different numerical integration functions did not affect the obtained results, however inaccuracy in numerical integration should be borne in mind as a source of potential error when implementing the routine with different programming languages.

The output values are written to the file energy.dat.

E.2.1 Octave Code for Heat Capacity Calculations

```
hbar = 1.055E-34;
tau_d = 0.49E-12; #Debye period
g_inf = 1.31E9; #High-f shear modulus
k = 1.38064852E-23; #Boltzmann constant
R = 8.314; #Ideal gas constant
omega_d = 2*pi/tau_d;
T = 0;
omega_f = 0;
no_rows = 71; #Number of rows in input data
T_interval = 5;
beta = 2.0E-4; #thermal expansion coeff.

sampleparam = zeros(no_rows, 2); #T, viscosity
energy = zeros (no_rows, 3); #Temp, Energy per particle, Cv

for i = 1:no_rows
  sampleparam(i,1) = 250 + T_interval*(i-1); #Calculate Ts
```

```
   #All viscosities need to be in micropascalsecs
   sampleparam(i,2) = …
   …(0.612E3)*exp(235.175/(sampleparam(i,1)-40.726));
   #Fit to viscosity
endfor

#Calculate E with existing values
for i = 1:no_rows #calculate energies
   energy(i,1) = sampleparam(i,1); #Copy temperature across
   T = sampleparam(i,1);
   omega_f = 2E6*pi*g_inf / sampleparam(i,2);
   #The 1E6 multip. is to get viscosity in correct units
   energy(i,2) = T*(1 + beta*T/2)*(3*debye(hbar*omega_d/(k*T)) …
   …- ((omega_f / omega_d).^3)*debye(hbar*omega_f/(k*T)));
endfor

for i = 2:no_rows #calculate heat capacities and errors
   energy(i,3) = (energy(i,2) - energy(i-1,2)) / T_interval;
endfor

#Save O/P file at the end
save energy.dat energy;

function retval = debye (x)

retval = (3/(x.^3)) * quadgk("debye_integrand", 0, x);

endfunction

function retval = debye_integrand (z)

retval = (z.^3)./(exp(z)-1);

endfunction
```

Bibliography

Undergraduate Level Textbooks

Fundamentals of classical and statistical thermodynamics, B.N. Roy, Wiley (2002).
Standard thermodynamics textbook, but includes coverage of Hildebrand mixing theory.

Gases, liquids and solids. D. Tabor, CUP (various editions from 1969).
The text gives a good but basic overview of the principal properties of the different states of matter. The description of the nature of van der Waals forces is particularly useful.

Heat and thermodynamics. J.K. Roberts and A.R. Miller, Blackie & Son (various editions).
Unfortunately out of print, this text provides extremely good revision of basic concepts of thermodynamics.

Introduction to the thermodynamics of materials, D.R. Gaskell, Taylor & Francis (2008).
Notable for four chapters covering miscibility and solutions in far more detail than is possible here.

Solid state physics, N.W. Ashcroft and N.D. Mermin, Harcourt (1976).
Key text on condensed matter physics, with an appendix containing the derivation of the dynamic structure factor $S(\mathbf{Q}, \omega)$.

Solid state physics, J.R. Hook and H.E. Hall, Wiley (1998).
Text giving good descriptions of concepts from solid state physics that are also applicable to fluids, for instance some aspects of diffraction theory.

Statistical physics, L.D. Landau and E.M. Lifshitz, Pergamon Press (1969).
Classic and comprehensive text on statistical physics.

X-ray diffraction, B.E. Warren, Addison-Wesley (1969).
Standard X-ray diffraction textbook but including detailed specific coverage of diffraction from fluids and amorphous solids.

Postgraduate Level Textbooks

High-pressure fluid phase equilibria. U.K. Deiters and T. Kraska, Elsevier (2012).
Authoritative textbook on thermodynamics of fluid mixtures including algorithms for computation of fluid mixture properties.

Kinetic theory of liquids. Y.I. Frenkel, Dover (1955).

Classic text by Frenkel which has stood the test of time, laying out the proposal that the liquid state close to melting curve should be understood by analogy to the solid state, and proposing the parameter of the liquid relaxation time.

Theory of simple liquids. J.-P. Hansen and I.R. McDonald, Academic Press (2006).

Extremely detailed textbook outlining the basic mathematical theory of liquids.

Papers and Theses

A new equation of state and tables of thermodynamic properties for methane covering the range from the melting line to 625 K at pressures up to 1000 MPa. U. Setzmann and W. Wagner, J. Phys. Chem. Ref. Data **20**, 1061 (1991).

Publication outlining the "Fundamental Equation of State" approach with good description of the underlying physics and experimental basis for the approach. Computational details are covered elsewhere.

Applications of supercritical carbon dioxide in materials processing and synthesis. X. Zhang, S. Heinonen and E. Levänen, RSC Advances **4**, 61137 (2014).

Sixteen-page review paper on this important topic.

Bond orientational ordering in liquids: Towards a unified description of water-like anomalies, liquid-liquid transition, glass transition, and crystallization. H. Tanaka, European Physical Letters **35**, 113 (2012).

Detailed (85 pages) review paper on bond orientational order in liquids with a focus on water, providing a more in-depth treatment than is possible in this text.

Brillouin scattering at high pressure: An overview. A. Polian, J. Raman Spec. **34**, 633 (2003).

Useful overview of the capabilities of Brillouin spectroscopy at high pressure in the diamond anvil cell, for the study of both solids and fluids.

Computer simulation as a tool for the interpretation of total scattering data from glasses and liquids. A.K. Soper, Mol. Sim. **38**, 1171 (2012).

Description of the empirical potential structure refinement (EPSR) method for obtaining liquid and glass structures from diffraction data including detailed information on two specific examples (water and dimethylsulphoxide).

Demonstration of the Allam Cycle: An update on the development status of a high efficiency supercritical carbon dioxide power process employing full carbon capture, R. Allam et al., Energy Procedia **114**, 5948 (2017).

A 19-page document describing development work to date and future prospects for supercritical CO_2 power generation based on the Allam cycle with built-in carbon capture and storage.

Is water one liquid or two? A.K. Soper, J. Chem. Phys. **150**, 234503 (2019).

Current thinking on the liquid-liquid phase transition in H_2O.

Melting curve equations at high pressure, V.V. Kechin, Phys. Rev. B. **65**, 052102 (2001).

Description of the rationale behind the Kechin equation, its relation to first principles and to other melting curve equations.

On the thermodynamics properties of fluids. E.H. Brown, Intnl. Inst. Refrig., Paris, Annexe **1960-1961**, 169 (1960).

Original (hard to obtain) publication setting out the theoretical basis for the existence of one
zeroth order inversion curve (the Zeno curve) and three first order inversion curves
(the Boyle curve, Amagat curve, and Joule-Thomson curve) in the condensing fluid.

*Quantitative structure factor and density measurements of high-pressure fluids in dia-
mond anvil cells by x-ray diffraction: Argon and water.* J. H. Eggert, G. Weck, P.
Loubeyre and M. Mezouar, Phys. Rev. B **65**, 174105 (2002).
Definitive paper on the methodology for obtaining the structure factor, radial distribution function, and density of fluids from X-ray diffraction data. The paper clearly
outlines what is possible, as well as the limitations of this method.

The characteristic curves of water. A. Neumaier and U.K. Deiters, Int. J. Thermophys. **37**,
96 (2016).
In-depth discussion of the zeroth and first order inversion curves, both in general and
with reference to water. An alternate unified method to define the curves is presented, and a number of additional mathematical conditions that are satisfied on
each of the curves are given.

The high-frequency dynamics of liquids and supercritical fluids. Ph.D. thesis of Filippo
Bencivenga (2006). https://tel.archives-ouvertes.fr/tel-00121509/ The thesis contains
introductory chapters giving a detailed review of structural and thermal relaxation
processes in fluids, focusing both on the mathematics and the underlying physics.

The many faces of packed column supercritical fluid chromatography: A review. E. Lesellier
and C. West, Journal of Chromatography A **1382**, 2 (2015).
A 45-page review paper on chromatography applications of supercritical CO_2.

Two liquid states of matter: A dynamic line on a phase diagram. V.V. Brazhkin, Yu.D.
Fomin, A.G. Lyapin, V.N. Ryzhov and K. Trachenko, Phys. Rev. E **85**, 031203 (2012).
Original paper laying out the theoretical basis for the Frenkel line ca. 60 years after the
original proposal of the liquid relaxation time parameter in *Kinetic theory of
liquids,* and related papers by Frenkel. See also V.V. Brazhkin et al., Phys. Rev. Lett.
111, 145901 (2013).

Free Software, Data, and Online Resources

The NIST webbook (REFPROP) (http://webbook.nist.gov/chemistry/) is an online tool
providing EOS data for key fluids as well as data on various parameters obtained
from the fundamental EOS. What is provided is not experimental data but the output from a fundamental equation of state for the fluid in question that has been
carefully selected. NIST only provides output for P, T conditions for which the fundamental equation of state is adequately backed by experimental data. This is why,
both here and elsewhere in the literature, the output from NIST is referred to as
experimental data even though that is not technically the case. In addition, NIST
lists the values of a number of useful properties for each fluid such as the critical
parameters. Software for use offline is also available, although not free of charge.

The ThermoC code (available from http://thermoc.uni-koeln.de/ to run online or down-
load) is an invaluable resource allowing the user to plot the EOS and various
parameters obtained from it for many fluids and fluid mixtures using different

EOS. Compared to NIST, the user has greater freedom to choose different EOS for a given fluid and to extrapolate beyond the ρ, T at which the EOS is backed by experimental data – at your own risk. The code is described in U.K. Deiters, Chem. Eng. Technol. **23**, 581 (2000).

Gudrun (available from https://www.isis.stfc.ac.uk/Pages/Gudrun.aspx) Software for analysis of raw neutron and X-ray diffraction data on fluids and glasses to obtain the differential scattering cross section. Manual contains detailed textbook-style introduction to diffraction from fluids and glasses, covering both theoretical and experimental aspects.

Empirical potential structure refinement (EPSR) software (available from https://www. isis.stfc.ac.uk/Pages/Empirical-Potential-Structure-Refinement.aspx).   Software to model the structure of fluids and glasses and refine the model by comparison to experimental $S(Q)$ data. Roughly speaking, the equivalent for fluids to the structural refinement software available for solids, e.g., GSAS, Topas, etc. Tutorial examples are included.

Dissolve software (available from https://github.com/trisyoungs/dissolve) and described in T. Youngs, Mol. Phys. **117**, 3464 (2019). Software currently under development, intended as long-term replacement for the EPSR code, compatible with modern diffraction instrumentation and combining refinement with MD capability. The software is fully integrated into a graphical user interface.

Jmol (available from http://jmol.sourceforge.net/). Software for modelling (intra-)molecular structure. Suitable for producing graphics/figures and also provides output in a form suitable for input into applications such as EPSR software.

Index

A

Adiabatic process 104–106, 123–127, 136, 238
  planetary adiabats 186, 192–194
  relation to isothermal process 238
Adiabatic to isothermal transition, *See* Positive sound dispersion
Allam cycle, *See* Thermodynamic cycles for power generation
Amagat curve 42, 127–134, 136, 211–233
  high temperature endpoint 132–133
  low temperature endpoint 134
  mathematical conditions 128–129, 131
  relation to second virial coefficient and pair potential 132–133
Ammonia 191–193
Argon (Ar)
  cohesive energy density 164
  Frenkel line 142–143, 208–209
  inversion curves 124–125, 127–128, 130–131, 136–138, 208–209
  melting curve 208–209
  P,T phase diagrams 75, 118, 124, 127, 131, 137, 143, 208–209
  phase change properties 48, 200
  ρ,T phase diagram 208–209
  vapour pressure curve 19, 24, 124, 127, 131, 208–209
  Widom lines 113–114, 116–119, 121–122, 208–209

B

Ballistic motion 10
Bank of terms, *See* Helmholtz function
Batschinski curve, *See* Zeno curve
Berthelot and Berthelot-Lorentz mixing rules 161–162
Boiling, *See* Vapour pressure curve
Boltzmann factor 62
Boron oxide (B_2O_3) 96
Bose-Einstein distribution 94
Boyle curve 42, 127–128, 130–138
  high temperature endpoint 132–133
  low temperature endpoint 134
  mathematical conditions 128, 130–131
  relation to second virial coefficient and pair potential 132–133
Brayton cycle, *See* Thermodynamic cycles for power generation
Brillouin scattering/spectroscopy 58, 79, 82, 96–98, 101, 206–207
  comparison to Raman spectroscopy 97
  experimental apparatus, scattering geometry 97
  frequency domain versus time domain 82
  Stokes versus anti-Stokes 97
Brillouin zone 98
Brown's conditions 42, 239–242
Bulk modulus 28, 47–51, 211–213, 220, 225–226
  pressure derivative 50–51, 212–213

C

Carbon dioxide (CO_2) 32, 88, *See also* Chromatography using supercritical fluids; Cleaning using supercritical fluids; Crystal and nanoparticle growth using supercritical fluids; Decaffeination; Drying using supercritical fluids; Exfoliation of layered materials using supercritical fluids; Thermodynamic cycles for power generation (carbon dioxide)
  capture and storage 171–173, 184–185
  Frenkel line 46, 167, 218, 220
  inversion curves 136, 217–218
  melting curve 217–220
  other food processing applications 175
  P,T phase diagrams 117–118, 167, 172–173, 217–220
  phase change properties 212–213
  planetary environments 181, 184–185, 187–189, 191
  ρ,T phase diagram 220–221
  short range order in fluid 220
  V,P phase diagram 172
  vapour pressure curve 217–218, 220–221
  Widom lines 117–119
Carbon nanotube suspensions in fluids 165
Carnahan-Starling EOS, *See* Equation of state
Carnot cycle, *See* Thermodynamic cycles for power generation
Characteristic curves, *See* Amagat curve; Boyle curve; Inversion curves (general comments); Joule-Thomson effect/curve
Charles curve, *See* Joule-Thomson effect/curve
Chromatography using supercritical fluids 176, 265
Classical liquid conditions 15–16, 65, 119, 199, 204–205, 221, 225
Clausius-Clapeyron equation 19–20, 69–70, 120
Cleaning using supercritical fluids 175–176
Co-ordination number 66–67
  behaviour at Frenkel line 144–146, 205–206
Cohesive energy density 160
  relation to real pressure 164–165
  relation to solubility parameter 160
  tabulated data for pure fluids 164
Collision broadening/narrowing, *See* Raman scattering/spectroscopy
Compressibility equation 63–65

Compression factor 40, 127–132, 134–136, 214, 240–242, *See also* Critical compression factor
Condensation, *See* Vapour pressure curve
Convection 191
Corresponding states principle 120, 199, 205, 226
Covalent bonding 17–18, 91
Critical compression factor 27, 32, 34
Critical isochore 114, 118–120, 141–143, 147–149, 151, 170, 175, 201, 205, 213–214, 227
Critical point 10, 15, 17, 19–27, *See also* Critical compression factor; Melting curve (Critical point)
  behaviour of cubic EOS 32, 34
  behaviour of fundamental EOS 39–41
  behaviour of van der Waals EOS 25–27
  compressibility at 31
  density at 18, 147–148
  ferromagnetic systems 21–23
  heat capacity divergence 22–24, 111–114
  opalescence 21
  temperature 20–21, 120
Cryogenic diamond anvil cell loading, *See* Diamond anvil cell
Crystal and nanoparticle growth using supercritical fluids 176–177
Cubic EOS, *See* Equation of state

D

de Broglie wavelength
  fluid particles 15, 204, 221, 225
  neutrons 53–55, 98
Debye
  frequency/period 12, 83, 86–87, 140–141, 260
  function 84, 260
  scattering equation 55–56
Decaffeination 175
Degrees of freedom 4–5, 82, 87, 141, 214, 223, 232
  rotational 4–5, 62, 79, 93
  translational 141
  vibrational 4–5, 79, 82, 232
Density of states 12–13, 83, 107
Departure function 162–163
Deuterium (D_2) 55, 220–221
Diamond anvil cell, *See also* Laser speckle method; Melting curve
  design 245–247
  double-stage 247

laser heating 150, 202–203, 206–207, 217, 219–220, 251–252
loading of fluids and fluid mixture samples 163–164, 247–249
maximum pressure 246–247
neutron scattering in 245
pressure measurement in 252–254
resistive heating 206–208, 215–217, 219–220, 223–224, 226–229, 232–233, 247, 249–251
Diffraction 3–4, 51–66, 68–69, 101, 121, 142, 144–146, 211–212, 226, 228, 232, 242–244, *See also* Debye (scattering equation); Radial distribution function $g(r)$; Structure factor $S(Q)$
atomic form factor $f(Q)$/scattering length b 52–55, 57
EPSR and other Monte Carlo methods 52, 61–64, 68–69
Fourier transform method in 52, 55–61, 242–244
Frenkel line 142, 144–146, 228, 232
isotopic substitution 55
Lorch modification 58–61
neutron/X-ray comparison 53–55
Q_{max}-cutoff 52, 57–58
scattering vector **Q** 53
solids 211–212, 226
spot size 55
subtraction of background/incoherent scattering 52
use in EOS measurement 52, 58–59
Diffusive motion of atoms in fluid 10, 73, 85, 138–139, 167, 176, 202
Dissociation
formation of plasma 4–6, 73, 150
of inter-atomic bonds 4–6, 150, 191, 220, 224, 232–233
Distinguishable particles 11, 16
Drying using supercritical fluids 175–176
Dwarf planets 182–190
Dynamic structure factor $S(\mathbf{Q}, \omega)$ *See* Inelastic neutron/X-ray scattering

E

Earth 181–182, 184–187, 189–190, 192
Empirical potential structure refinement, *See* Diffraction
Energy equation 63–65, 159

Enthalpy (H) 36–37, 39, 238, 123–125, 130–131, 136, 238, *See also* Adiabatic process
ideal gas 37, 238
relation to Helmholtz function 36, 39–40
Entropy 18–20, 36–37, 39–40, 65, 157–160, 162–163, 237–239
change on mixing 157–160, 162–163
ideal gas 36–39, 65
relation to $g(r)$ 65
relation to Helmholtz function 36, 39–40, 162–163
system in equilibrium 238
Equation of state (EOS), *See also* Miscibility/mixtures of fluids
availability and implementation 41, 44–46, 255–259
Carnahan-Starling 31, 35–36, 92–93, 147
cubic (general comments) 23, 27–35, 40–44, 134, 255–257
extrapolation 32, 45, 72, 136–138, 141–142, 171, 173, 205, 211–215, 217, 220, 225–228, 231–233, 265
fundamental/reference 11–12, 23, 31, 36–51, 59, 87–88, 112–116, 120, 124–125, 127, 131, 135–138, 141, 143, 146, 151, 162–163, 170, 172–173, 199–202, 204–206, 208–209, 211–223, 225–228, 231–233, 260
GERG-2008 162–163
hard sphere 31, 35–36, 91–93, 147
IAPWS 170, 211, 226–230
ideal gas 1–5, 10, 25, 42, 128, 156
invertibility 32, 34, 50, 69
Murnaghan 50–51, 211–213, 217, 228, 230
Patel-Teja 34, 41–45
Peng-Robinson 34–35, 41–44, 136–138
planetary environments 188
prediction of inversion curves 42, 136–138
prediction of vapour pressure curves 28–29, 34–35, 42–43
Redlich-Kwong 34
Redlich-Kwong-Soave 34, 41–44, 136–138
validity/testing 25–29, 39–44
van der Waals 1–3, 24–29, 33–34, 111–112
virial 1–2, 32, 35, 40, 132–133, 161
volume translation 34–35
Xiang-Deiters 45, 205–206
Ethane (C_2H_6) 67, 248
EOS 42, 45
planetary environments 182
Europa 182

Exfoliation of layered materials using
supercritical fluids 177

F

Ferromagnetic system 21–23
First law of thermodynamics 237–238
First order inversion curves, *See* Amagat
curve; Boyle curve; Joule-Thomson
effect/curve; Inversion curves
(general comments)
First order phase transition 20, 30, 70–72, 111,
138, 150, 166, 171, 215
Fisher-Widom line 121–123
Fourier transform, *See* Diffraction; Inelastic
neutron/X-ray scattering
Free energy, *See* Enthalpy; Gibbs function;
Helmholtz function
Frenkel line 94, 111, 138–151, *See also* Argon
(Ar); Carbon dioxide (CO_2); Helium
(He); Hydrogen (H_2/H); Methane
(CH_4); Neon (Ne); Nitrogen (N_2);
Water (H_2O)
comparison to Widom lines 147–149
definitions 138–146
experimental measurement 140–146
path on the *P,T* phase diagram 141–143,
150–151
path on the ρ,*T* phase diagram 147–149
planetary environments 191, 194
positive sound dispersion at 148
relation to co-ordination number 144–146
relation to heat capacity 140–142
relation to intramolecular vibrations
142–144
relation to mesoscopic static structure 140
relation to shear wave propagation 140–142
relation to velocity autocorrelation
function and relaxation time
138–140
termination 150
Frenkel relaxation time, *See* Relaxation time
Fusion, *See* Melting curve
Fundamental EOS, *See* Equation of state

G

Gas, *See also* Equation of state; Radial
distribution function *g(r)*; Structure
factor *S(Q)*

ideal 1–5, 10, 25, 36–39, 42, 63, 65, 125, 128,
134, 156, 162, 238, 240–241
miscibility 1, 153, 166
real/non-ideal 1–10, 12, 17–20, 23–29, 51,
63–65, 72, 88, 90–91, 94, 104, 111–112,
125, 126–129, 138–139, 144–147, 153,
162, 166–167, 160–174
Generalized hydrodynamics (definition) 12
GERG-2008, *See* Equation of state
Gibbs function
in boiling/condensation transition 17,
19–20, 28–29
change on mixing 156–159, 239, 255–256
in melting/crystallization transition 69
relation to Helmholtz function 40
Grüneisen parameter 12, 83–87, *See also*
Raman scattering/spectroscopy
hard sphere fluid models 35–36, 67, 91–93
Lennard-Jones potential 91
relation to thermal expansion coefficient 83

H

Hard sphere EOS, *See* Equation of state
Heat capacity, *See also* Frenkel line (relation
to heat capacity); Widom line (heat
capacities)
change at vapour pressure curve/critical
point 22–24, 113–114, 208, 231
effect of intramolecular bonds breaking
5–6
effect of intramolecular degrees of freedom
5–6, 37, 141, 214, 223, 232
ferromagnetic system 22–23
fluid (general) 82–88, 260
ideal gas 4, 37
real gas 4–6
relation to Helmholtz function 37, 40
relation to Joule-Thomson coefficient 124–125
relation to internal energy 4, 82–83
Helium (He) 15–16, 119, 199–205, *See also*
Miscibility/mixtures of fluids
(Hydrogen and Helium)
planetary environments 191–192
Helmholtz function (*F*)
bank of terms 39
ideal gas component 36–39
at inversion curves 131, 136
for mixtures 162–163
relation to experimental observables 11–12,
39–41, 131

relation to internal energy 40, 82–83
relation to isothermal changes 238–239
relation to partition function 11–12, 16
reduced (definition) 38
residual component 39–40, 120
self-consistency 39
separation into kinetic and potential
    components using classical liquid
    approximation 15–16
Henry's law, *See* Miscibility/mixtures of fluids
Hildebrand theory, *See* Miscibility/mixtures
    of fluids
Hydrodynamics (definition) 12
Hydrogen (H_2/H) 5–6, 15, 220–226, *See*
    also Deuterium (D_2); Diffraction
    (isotopic substitution); Miscibility/
    mixtures of fluids (Hydrogen and
    Helium); Tritium (T_2)
    Frenkel line 194
    inversion curves 221–222
    melting curve 73, 221–225
    metallic phase(s) 191
    P,T phase diagrams 221–223
    phase change properties 211–212
    planetary environments 150, 191–192, 194
    Raman spectra 6, 93
    ρ,T phase diagram 224–226
    vapour pressure curve 221–222
    Widom lines 119, 221–222
Hydrogen bonding 18, 69, 119, 166
Hydrophobicity, *See* Miscibility/mixtures of
    fluids

I

IAPWS EOS, *See* Equation of state
Ideal fluid (as opposed to condensing fluid)
    239–242
Ideal gas, *See* Gas
Incoherent scattering, *See* Diffraction
Indistinguishable particles 11, 16
Inelastic neutron/X-ray scattering 79, 82, 98–106,
    See also Positive sound dispersion
    dynamic structure factor 101–102
    use to obtain dispersion relations 98–106
Infinite frequency shear modulus, *See* Shear
    modulus
Internal energy, *See also* Energy equation; Heat
    capacity; Helmholtz function (*F*);
    Miscibility/mixtures of fluids (change
    in internal energy on mixing)

at Amagat curve 129–130
contributions from longitudinal
    and shear waves 12–13,
    82–85, 107, 140–141
ideal gas 4
in Joule-Thomson effect 123
relation to other thermodynamic
    functions 237–239
Intramolecular bond 6–7, 17, 58, 62, 67,
    87–88, 92–93, 142, 145, 165, 214,
    223, 232
Inversion curves (general comments), *See*
    also Amagat curve; Boyle curve;
    Joule-Thomson effect/curve; Zeno
    curve
    definitions in terms of compression factor
    128, 131–134
    definitions in terms of Helmholtz
    function 131
    high temperature endpoint 132–134
    mathematical conditions (summary) 131
    nomenclature 131
    relation to second virial coefficient and
    pair potential 132–134
    unified approach to inversion curves
    127–138
    use to verify EOS 136–138
Iron (Fe) fluid 190, 194
Isenthalpic process, *See* Adiabatic process
Isotopic substitution, *See* Diffraction

J

Joule curve, *See* Amagat curve
Joule-Kelvin effect, *See* Joule-Thomson effect/
    curve
Joule-Thomson effect/curve
    balance between attractive and repulsive
    forces 126–127
    coefficient 124–126
    high temperature endpoint 124–127,
    132–134
    isenthalpic/adiabatic nature 123–124
    low temperature endpoint 133–134
    relation to critical point 125–126
    relation to second virial coefficient and
    pair potential 132–134
    use to cool fluids 123–125
    van der Waals fluid 125–126
Jupiter 182, 190–194
Juno spacecraft 192

K

k-gap 107
Kechin equation, *See* Melting curve

L

Laminar flow 8
Laser speckle method, *See* Melting curve
Lennard-Jones fluid 17–18, 41, 62–65, 83, 91, 121, 123, 132–133, 139–140, 147, 149–150
Lennard-Jones potential, *See* Pair potential
Lennard-Jones units 140
Lever rule, *See* Miscibility/mixtures of fluids
Liquid relaxation time, *See* Relaxation time (Frenkel)
Liquid-liquid phase transitions 67–69, 200–201, 220, 227–228, *See also* Frenkel line
Long-range order/lattice 1, 3, 10, 12, 72–73, 97, 121
Longitudinal (sound) waves 12–13, 18, 40, 82–88, 96–98, 102–106, 112–113, 116–119, 141
Lorch modification, *See* Diffraction
Lorentz mixing rule 162

M

Mandl, F. 72
Mars 182, 187, 189
Matlab 255
Maxwell
    construction 28–29, 255–257
    relations xvi, 125
    velocity distribution 8–9
    viscoelastic liquid, relaxation time 12, 79–81, 107
Mean free path 8–9
Melting curve
    concavity towards temperature axis 71–72, 219–220
    critical point 72–74
    density change 12, 42, 45, 47–50, 69–70, 72–74, 200, 212–213
    glatzel equation 69, 74, 76
    gradient 69–74, 219–220, 223–224
    Kechin equation 69–72, 202–203, 222–224, 232–233

    measurement using laser speckle method 203, 206–207
    Simon-Glatzel equation 69–72, 74–75, 200, 206–208, 215–216, 219–220, 232–233
    temperature maximum 72–73, 222–224
    termination 72–74
    two-phase and one-phase approaches 69–70
Mercury 182, 187
Methane (CH_4)
    EOS 32, 37, 42–45, 217
    Frenkel line 141–142, 144, 146, 151
    inversion curves 134–136, 214–215
    melting curve 215–216
    orientational order in fluid and solid states 211–212
    P,T phase diagrams 214–216
    phase change properties 212–213
    planetary environments 182, 192–193
    ρ,T phase diagram 215–216
    Raman spectrum 6–7, 94–95
    vapour pressure curve 214–215
    Widom lines 114–115, 117–119, 213–214
    X-ray diffraction 53
Minerals
    dissolution 182
    reactions 184, 189
    solubility in water 165–166
Miscibility/mixtures of fluids, *See also* Cohesive energy density; Equations of state (GERG-2008); Gases (miscibility); Helmholtz function (for mixtures)
    change in internal energy on mixing 157–160
    above the critical temperature 166–167
    entropy change on mixing 157, 159–160, 162–163
    EOS for mixtures 160–163
    Gibbs function of Raoultian mixtures 156–158
    Henry's law 155–156
    Hildebrand theory 158–162
    hydrocarbons and water 153
    hydrogen and helium 192–193
    lever rule 158
    MD simulations 164–167
    mixing rules 161–162
    phase equilibria in miscible fluids 148
    planetary environments 192–193

preparation in the diamond anvil cell 163–164, 249
Raman spectra of fluid mixtures 164–165
random mixing approximation 161
Raoult's law 154–158
treatment with fundamental EOS 162–163
water, carbon dioxide and NaCl 184
Miscibility/mixtures of solids (solid solutions) 153
Mixing rules, *See* Miscibility/mixtures of fluids
Molecular dynamics (MD) 10–11, 49, 62, 148, 214, 232, 266, *See also* Miscibility/mixtures of fluids; Raman scattering/spectroscopy; Velocity autocorrelation function
Momentum
conservation in inelastic scattering 97–100
of gas particles 9–10
position-momentum uncertainty principle 73
Monte Carlo simulation, *See* Diffraction
Murnaghan EOS, *See* Equation of state

N

Neon (Ne)
EOS 149, 205–208
Frenkel line 141–142, 144–149
inversion curves 205–206
melting curve 206–207
P,T phase diagrams 205–207
phase change properties 200
positive sound dispersion 104, 106
ρ,T phase diagram 206, 208
vapour pressure curve 205–206
Widom lines 119, 149, 205
Neptune 182, 191–193
Neutron diffraction, *See* Diffraction
NIST REFPROP, *See* REFPROP
Nitrogen (N_2), *See also* Thermodynamic cycles for power generation (nitrogen)
in atmosphere of Venus 188–189
co-ordination number in fluid 67, 144–146
EOS 49–51
fluid $S(Q)$ compared to EPSR simulations 63–64
Frenkel line 144–146, 231–233
heat capacity 4–5, 145–146, 231–233
inversion curves 231–232
melting curve 73–74, 76, 231–233

P,T phase diagrams 231–233
pair potential 17–18
phase change properties 49–51, 212–213
planetary interiors 182, 188–189
ρ,T phase diagram 232–233
Raman spectrum 91–92, 94–95
simulated $g(r)$ using EPSR 66
speed of molecules in air at 300 K 9
structure factor $S(Q)$ in gas state 51
vapour pressure curve 231–233
Widom lines 119, 231–233
Noble gas fluid (general comments) 17, 63, 83, 87, 141
Non-hydrostatic stress 94, 142, 247

O

Octave 255–261
Onsager, L. 23
Order, *See* Long-range order/lattice; Short-range order in fluids; Orientational order in fluids
Orientational order in fluids 3, 49, 67–69, 211–213, 220, *See also* Carbon dioxide (CO_2); Methane (CH_4); Water (H_2O)
Oxygen (O_2) 102–103, *See also* Thermodynamic cycles for power generation (Allam cycle)

P

Pair distribution function, *See* Radial distribution function
Pair potential 2, 8–10, 17–18, 35–36, 62–65, 83, 90–92, 120–123, 132–133, 147, 159–162, *See also* Lennard-Jones fluid
Lennard-Jones for fluid mixtures 159–162
Lennard-Jones for pure fluids 8–10, 17–18, 62–65, 83, 90–91, 120–123, 132–133, 147
Partition function
ferromagnetic system 21–23
fluid 11, 15–16, 21, 36, 63–65
Patel-Teja EOS, *See* Equation of state
Peng-Robinson EOS, *See* Equation of state
Phonons, *See also* Longitudinal (sound) waves; Transverse/shear waves
fluids 12–13, 79–88, 96–98, 101–106
solids 10, 47

Planetary atmospheres 188–189

Planetary interiors 31, 150, 182, 185–186, 190–192, 221

Plasma 5–6, 73, 150

Pluto 182, 190

Positive sound dispersion 103–106, 148

Pressure (relation to Helmholtz function) 11, 40

Pressure equation 63–65, 160–161

Q

Q_{max}-cutoff, *See* Diffraction

Quantum liquid 15–16, 199–201, 204–205, 220–221, 225

Quantum states 4, 16, 93

Quantum well 20–21

Quartz 177, 182–183, 189

R

Radial distribution function $g(r)$ 3, 19, 52, 55–66, 121–123, 146, *See also* Diffraction

Raman scattering/spectroscopy, *See also* Miscibility/mixtures of fluids

collision broadening/narrowing 94, 142

comparison to Brillouin spectroscopy 97

comparison to inelastic neutron/X-ray scattering 250

dense fluids 88–96, 142

in the diamond anvil cell 250–252

effect of pressure/density change in gases 6–7

effect of temperature change in gases 6–7

Frenkel line 142, 144, 146, 232

instrumental spectral resolution 6, 94

intensity and linewidth 93–96, 142, 144

melting curves 219, 232

observation of liquid-liquid transition in H_2O 228

prediction of Raman spectra using MD 94, 96

role of Grüneisen parameter(s) 90–91, 142

rotational 93

solids 6, 88

Stokes versus anti-Stokes peaks 94

triple-grating versus single-grating spectrometers 93

units for frequency 6

Random mixing approximation, *See* Miscibility/mixtures of fluids

Rankine cycle, *See* Thermodynamic cycles for power generation

Raoult's law, *See* Miscibility/mixtures of fluids

Rayleigh scattering 93, 97

Redlich-Kwong EOS and Redlich-Kwong-Soave EOS, *See* Equation of state

Reduced Helmholtz function, *See* Helmholtz function

Reduced pressure, temperature, volume 24–27, 33, 38, 48–50, 119–123, 126, 140, 163, 199, 205

Reference EOS, *See* Equation of state

REFPROP xii, 44–45, 48–51, 265

Refractive index of fluid 58

Relaxation time (Frenkel) 12, 81–82, 84–88, 94, 107, 138–142, 150, *See also* Maxwell (relaxation time);

Relation to positive sound dispersion/ viscous-elastic transition 148

Residual Helmholtz function, *See* Helmholtz function

Rotational degrees of freedom, *See* Degrees of freedom

Ruby photoluminescence, *See* Diamond anvil cell

S

Saturated liquid 25, 34, 200, 204, 206, 208, 212–213, 220–221, 224–225, 232–233

Saturated vapour 4, 25, 34, 200, 212–213

Scattering vector Q, *See* Diffraction

Second law of thermodynamics 20, 237

Second virial coefficient $B_2(T)$, *See* Virial coefficients

Second order phase transition 20, 111–112, 120

Shear modulus 81

infinite frequency G_∞ 86–88, 260

Shear waves, *See* Transverse/shear waves

Short-range order in fluids 49, 67–69, 72–74, 121, *See also* Carbon dioxide (CO_2); Methane (CH_4); Water (H_2O)

Single reheat cycle, *See* Thermodynamic cycles for power generation

Solids, *See also* Equation of state (Murnaghan); Raman scattering/ spectroscopy;

Orientational (dis)order 49

Solubility of solids in fluids 155, 160, 165–167, 175–176, 183–184, 187

Solubility parameter, *See* Cohesive energy density
Sound, *See* Longitudinal (sound) waves; Positive sound dispersion; Speed of sound; Transverse/shear waves
Specific heat capacity, *See* Heat capacity
Speed of sound 18, 107, 113, 116–118, 221
    relation to Helmholtz function 40
Spinodal curve 28–29
Stirling's formula 16
Structure factor $S(Q)$, *See also* Diffraction (EPSR and other Monte Carlo methods); Inelastic neutron/X-ray scattering (dynamic structure factor)
    extraction from raw diffraction data 56–57
    Fourier transform to obtain $g(r)$ 55–61
    gas 3–4, 51
    general comments 19, 51–52, 244
Subduction 182, 187
Sublimation 213
Surface energy 21, 28, 177
Surface tension 175

T

Talc 184
Thermal conductivity 10, 113, 117–118, 199, 221, 226
Thermal expansion coefficient 12, 42, 83, 87, 228, 260
    relation to Grüneisen parameter(s) 83, 87
ThermoC code 44–45, 114, 124–125, 131, 134–137, 151, 172–173, 201, 204, 206, 208–209, 214, 217, 220–221, 226, 231–233, 265
Thermodynamic cycles for power generation
    Allam cycle 171–173
    Brayton cycle 170–174
    carbon dioxide 171–173
    Carnot cycle 169
    crossing of Widom lines and Frenkel line 170–173
    fossil fuel power stations 170–173
    nitrogen 173–174
    nuclear power stations 170–171, 173–174
    Rankine cycle 169
    single reheat cycle 170
    water 170–171
Third virial coefficient, *See* Virial coefficients

Translational degrees of freedom, *See* Degrees of freedom
Transverse/shear waves
    direct experimental observation in fluids 96–98
    dispersion relation in k-gap model 107
    effect on fluid heat capacity 83–85, 140–142, 260
    general comments 12, 19, 81–82
    modelling using MD 11
    seismic s-waves 81
Triple point 42, 48–49, 69–70, 72, 74–76, 139, 200, 206–208, 211–212, 219–220, 223–225, 229, 232–233
Tritium (T_2) 220–221

U

Uncertainty principle 55, 73
Uranus 182, 191–193

V

van der Waals, *See also* Equation of state (van der Waals)
    fluid 111–112, 125
    forces/bonding 6, 17–18, 25, 63, 88, 91, 119, 142, 177
    loops 28–29, 44, 255–257
    radius 25, 64, 91–93, 147–148
Vapour pressure curve 4–5, 15, 17–20, 31–35, 42–44, 88, 111–112, 117, 120–121, 123–125, 133–135, 138–139, 142, 158, 171, 177, *See also* Argon (Ar); Carbon dioxide (CO_2); Clausius-Clapeyron equation; Critical point; Helium (He); Hydrogen (H_2/H); Methane (CH_4); Neon (Ne); Nitrogen (N_2); van der Waals (loops); Water (H_2O)
    as gas-gas transition close to critical point 19
VeGa 2 spacecraft 188–189
Velocity autocorrelation function 79, 138–140, 146, 148, 151, 167, 171–173, 200–201, 204, 214–215, 218, 220–221, 228–229
Venus 182, 187–189
Vibrational degrees of freedom, *See* Degrees of freedom
Virial coefficients 2, 35–36
    Second $B_2(T)$ 2, 40, 132–133
    Third $B_3(T)$ 40

Virial EOS, *See* Equation of state
Virial theorem 85
Viscosity
   gases 8–10
   liquids 80, 86–87, 98, 176, 199, 260
   at Widom line 111,113, 116–118, 121–122
Viscous to elastic transition, *See* Positive
      sound dispersion
Vogel-Fulcher-Tammann equation 87, 260

W

Waldman and Hagler mixing rule 161–162
Water (H_2O), *See also* Crystal and
      nanoparticle growth using
      supercritical fluids; Equation of
      state (IAPWS); Miscibility/mixtures
      of fluids (hydrocarbons and water);
      Thermodynamic cycles for power
      generation (water)
   cohesive energy density 165
   density anomalies 42, 47, 228–230
   diffraction measurements 57–59, 65, 67–69
   EOS near melting curve 228–230
   Frenkel line 171, 226–229
   inversion curves 134–136, 226–228
   liquid-liquid transition 68–69, 227–228
   melting curve 70, 226–229
   P,T phase diagrams 186, 226–229
   phase change properties 212–213
   planetary environments 181–187, 189–191,
      192–193
   ρ,T phase diagram 228–229
   vapour pressure curve 226–229
   Widom lines 117–119, 171, 226–227
Widom lines 111–121, *See also* Argon (Ar);
      Carbon dioxide (CO_2); Helium (He);
      Hydrogen (H_2/H); Methane (CH_4);
      Neon (Ne); Nitrogen (N_2); Water
      (H_2O)
   comparison to Frenkel line 147–148
   crossing in industrial processes 170–173

density/compressibility 31, 111–113, 116
divergence and termination 112–114,
      119–120
as function of reduced temperature 119–120
as a function of density 121–122
heat capacities 113, 116–117
lack of second order phase transition
      111–112
path on P,T phase diagram 117–119
path on ρ,T phase diagram 147–148
phenomenological fitting 113–117
relation to principle of corresponding
      states 119–120
relation to vapour pressure curve 31,
      111–112, 120–121
sound speed 117
van der Waals fluid 111–112
viscosity/thermal conductivity 113, 117,
      121–122

X

X-ray diffraction, *See* Diffraction

Z

Zeno curve 127–128, *See also* Argon (Ar);
      Carbon dioxide (CO_2); Helium (He);
      Hydrogen (H_2/H); Methane (CH_4);
      Neon (Ne); Nitrogen (N_2);
      Water (H_2O)
   high temperature endpoint 128, 130,
      132–133
   low temperature endpoint 133–136
   mathematical conditions 131
   path on P,T phase diagram 127, 131
   physical origin 128, 134
   relation to second virial coefficient and
      pair potential 132–133
Zero-point energy 73, 82
Zeroth order inversion curve, *See*
      Zeno curve

Printed in the United States
by Baker & Taylor Publisher Services